Lecture Notes in Computer Science 10378

Commenced Publication in 1973
Founding and Former Series Editors:
Gerhard Goos, Juris Hartmanis, and Jan van Leeuwen

More information about this series at http://www.springer.com/series/7408

Nigel Thomas · Matthew Forshaw (Eds.)

Analytical and Stochastic Modelling Techniques and Applications

24th International Conference, ASMTA 2017
Newcastle-upon-Tyne, UK, July 10–11, 2017
Proceedings

 Springer

Editors

Nigel Thomas
School of Computing Science
Newcastle University
Newcastle upon Tyne
UK

Matthew Forshaw
School of Computing Science
Newcastle University
Newcastle upon Tyne
UK

ISSN 0302-9743   ISSN 1611-3349   (electronic)
Lecture Notes in Computer Science
ISBN 978-3-319-61427-4   ISBN 978-3-319-61428-1   (eBook)
DOI 10.1007/978-3-319-61428-1

Library of Congress Control Number: 2017943839

LNCS Sublibrary: SL2 – Programming and Software Engineering

Printed on acid-free paper

This Springer imprint is published by Springer Nature
The registered company is Springer International Publishing AG
The registered company address is: Gewerbestrasse 11, 6330 Cham, Switzerland

Preface

This volume contains the papers presented at ASMTA 2017: the 24th International Conference on Analytical and Stochastic Modelling Techniques and Applications held during July 10–11, 2017, in Newcastle upon Tyne, UK.

Owing to the number of concurrent calls for papers in the field, the number of submissions was considerably smaller than previous years. There were 27 submissions. Each submission was reviewed by, on average, 3.4 Program Committee members. The committee decided to accept 14 papers.

This was the 24th year of ASMTA, which shows a considerable durability in a rapidly evolving field. Over the years ASMTA has been the forum for many important papers investigating the key topics of the day in the area of analytical and stochastic modelling. In this volume we are delighted to have contributions employing a diverse range of analysis techniques, including queueing theoretical results, stochastic Petri nets, proxel-based simulation, stochastic bounds, and reversible Markov chains. The range of topics within a small number of papers is impressive and demonstrates the power of stochastic analysis to tackle challenging problems in complex computer and communication systems.

We would like to take this opportunity to thank those who helped put ASMTA 2017 together, in particular Khalid Al-Begain, without whom ASMTA would not exist. Dieter Fiems was extremely helpful in passing on his experience and in managing the conference website. We would also like to thank our colleagues in Newcastle, Jen Wood and Claire Smith, who helped with practical arrangements and bookings, and our PhD students, who acted as a local support team during the conference. Finally we would like to acknowledge the continued support of Springer in publishing the proceedings and the team at EasyChair for providing comprehensive conference support with no charge.

May 2017

Nigel Thomas
Matthew Forshaw

Organization

Program Committee

Sergey Andreev	Tampere University of Technology, Finland
Jonatha Anselmi	Inria, France
Konstantin Avrachenkov	Inria Sophia Antipolis, France
Christel Baier	Technical University of Dresden, Germany
Simonetta Balsamo	Università Ca' Foscari di Venezia, Italy
Koen De Turck	Université Paris Saclay, France
Ioannis Dimitriou	University of Patras, Greece
Antonis Economou	University of Athens, Greece
Dieter Fiems	Ghent University, Belgium
Matthew Forshaw	Newcastle University, UK
Jean-Michel Fourneau	Université de Versailles St. Quentin, France
Marco Gribaudo	Politecnico di Milano, Italy
Yezekael Hayel	LIA/University of Avignon, France
András Horváth	University of Turin, Italy
Gábor Horváth	Budapest University of Technology and Economics, Hungary
Stella Kapodistria	Eindhoven University of Technology, The Netherlands
Helen Karatza	Aristotle University of Thessaloniki, Greece
William Knottenbelt	Imperial College London, UK
Lasse Leskelä	Aalto University, Finland
Daniele Manini	University of Turin, Italy
Andrea Marin	University of Venice, Italy
Yoni Nazarathy	The University of Queensland, Australia
Jose Nino-Mora	Carlos III University of Madrid, Spain
Antonio Pacheco	Instituto Superior Tecnico, Portugal
Tuan Phung-Duc	University of Tsukuba, Japan
Balakrishna Prabhu	LAAS-CNRS, France
Juan Pérez	Universidad del Rosario, Colombia
Marie-Ange Remiche	University of Namur, Belgium
Anne Remke	WWU Münster, Germany
Jacques Resing	Eindhoven University of Technology, The Netherlands
Marco Scarpa	University of Messina, Italy
Bruno Sericola	Inria, France
Devin Sezer	Middle East Technical University, Turkey
János Sztrik	University of Debrecen, Hungary
Miklos Telek	Budapest University of Technology and Economics, Hungary
Nigel Thomas	Newcastle University, UK

Dietmar Tutsch	University of Wuppertal, Germany
Jean-Marc Vincent	Laboratoire LIG, France
Sabine Wittevrongel	Ghent University, Belgium
Verena Wolf	Saarland University, Germany
Katinka Wolter	Freie Universität zu Berlin, Germany
Alexander Zeifman	Vologda State University, Russia

Additional Reviewers

Dei Rossi, Gian-Luca
Gardner, Kristen
Horvath, Illes

Contents

Stochastic Bounds for Switched Bernoulli Batch Arrivals Observed Through Measurements

Farah Aït-Salaht[1], Hind Castel-Taleb[2], Jean-Michel Fourneau[3], and Nihal Pekergin[4(✉)]

[1] LIP6, Ensai, Rennes, France
[2] SAMOVAR, UMR 5157, Télécom Sud Paris, Evry, France
[3] DAVID, UVSQ, Univ. Paris Saclay, Versailles, France
[4] LACL, Univ. Paris Est, Créteil, France
nihal.pekergin@u-pec.fr

Abstract. We generalise to non stationary traffics an approach that we have previously proposed to derive performance bounds of a queue under histogram-based input traffics. We use strong stochastic ordering to derive stochastic bounds on the queue length and the output traffic. These bounds are valid for transient distributions of these measures and also for the steady-state distributions when they exist. We provide some numerical techniques under arrivals modelled by a Switched Batch Bernoulli Process (SBBP). Unlike approximate methods, these bounds can be used to check if the Quality of Service constraints are satisfied or not. Our approach provides a tradeoff between the accuracy of results and the computational complexity and it is much faster than the histogram-based simulation proposed in the literature.

1 Introduction

Measurements and traces are now much more frequent and we advocate that we can use them to make the performance analysis of networking elements more precise and more realistic. Typically, the traces are used as an input for a fitting algorithm which finds the best approximation inside a class of well-known stochastic processes (see, for instance, [11]). When this process can be associated to a Markov process or chain, the whole system can be modelled by a so-called structured Markov chain (see [12] for an example) and many algorithms have been derived to solve the steady-state distribution for this type of models.

In [1,2], we have proposed a different approach for stationary arrivals: we model the system in discrete time and we use directly the measurements to obtain a discrete distribution of arrivals during a time slot. Thus, we avoid the fitting procedure and the approximations it may add in the model. Such an approximation due to the fitting of the processes may lead to incorrect results (see [5] for such a problem for service time distributions).

Such an idea has already been proposed and is known as the histogram based models for more than 20 years (see for instance, the work by Skelly et al. [15] in the area of network calculus to model the video sources and to predict

N. Thomas and M. Forshaw (Eds.): ASMTA 2017, LNCS 10378, pp. 1–15, 2017.
DOI: 10.1007/978-3-319-61428-1_1

buffer occupancy distributions). Recently, Hernàndez et al. [8–10] have introduced an approach called HBSP (Histogram Based Stochastic Process) to obtain histograms of buffer occupancy. They use histograms as inputs and some specific operators in discrete time to represent a finite capacity buffer with a constant service under the First Come First Served (FCFS) discipline. The model is solved numerically and as usual, the curse of dimensionality appears. When the number of bins in the histograms is too large, the computation times become extremely high and the authors present an approximation of the histograms of traffic which leads to a smaller complexity and a faster resolution. Unfortunately the accuracy of the approximation cannot be checked.

We propose a more accurate method to deal with histograms having a large number of bins. First in [3] we prove that the system is stochastically monotone. This allows to obtain bounds on the queue size and the output process when we consider bounds on the input process. Second, in [1] we provide several algorithms to derive stochastic bounds of the arrival process with a smaller complexity. As we build lower and upper bounds, our approach provides an estimation of the approximations. The complexity in the numerical computations in basically dependent of the number of bins in the histogram or the number of atoms in the discrete distribution. The main assumption of the approach is the stationarity of the input process.

Here, we do not assume that the traffic is stationary. Typical Internet services such as web surfing and high speed streaming services (Video On Demand (VOD) and video conferencing), tend to generate sporadic traffic, and hence it would be realistic to consider bursty packet arrivals for today's telecommunication traffic. There are some interesting queueing models and analytical results considering bursty sources and discrete time queueing systems.

In [18], they consider finite capacity queue in discrete time with constant service time of arbitrary length, and bursty on/off source with geometric distributed lengths of the phase. Closed form are derived for the loss ratio of cells. In [19] an infinite capacity discrete-time queue with Bernoulli bursty source and batch arrivals is analysed using the generating function technique. A closed form expressions of some performance measures as average buffer length, and average delays are obtained. Markov modulated arrivals have been quite often considered in the literature to represent traffic arrivals [4,14]. In [14], they define an MMPP (Markov Modulated Poisson Process) traffic model that accurately approximates the characteristics of Internet traffic traces. Results prove that the queuing behaviour of the traffic generated by the MMPP model is coherent with the one produced by real traces. Some important results on MMPP traffic and queues with MMPP input are described in [4].

In this paper, we propose to apply stochastic bounds on the input traffic to derive stochastic bounds on the queue length and the departure flow. We propose a numerical technique to compute the bounds in an efficient way. We show how our approach which has been developed for stationary arrivals can be generalised to Switched Bernoulli Batch Process (SBBP in the following).

The technical part of the paper is as follows. We introduce briefly bounds for the $\leq_{st}$ ordering in the next section for the sake of completeness. We advocate that monotonicity of the evolution equation as well as stochastic bounds may help to solve such a queueing model when the arrival process is not stationary. We first considered the stationarity assumption to derive some results, theorems and algorithms in Sect. 3 which will be then generalised for non stationary arrival processes in Sect. 4.

2   A Brief Presentation of Stochastic Comparison

We refer to [13] for theoretical issues of the stochastic comparison method. We consider state space $\mathcal{G} = \{1, 2, \ldots, n\}$ endowed with a total order denoted as $\leq$. Let X and Y be two discrete random variables taking values on $\mathcal{G}$, with cumulative probability distributions F_X and F_Y, and probability mass functions (pmf) $d2$ and $d1$. The ith index of pmf vectors denotes the probability that the underlying random value takes value i: $d2(i) = \text{Prob}(X = i)$, and $d1(i) = \text{Prob}(Y = i)$, for $i = 1, 2, \ldots, n$. The stochastic comparison of two random variables in the sense of the strong stochastic order, $\leq_{st}$ can be defined as follows.

Definition 1. *The following definitions are equivalent.*

- **generic definition:**

$$X \leq_{st} Y \Longleftrightarrow \mathbb{E}f(X) \leq \mathbb{E}f(Y),$$

for all increasing (non decreasing) functions $f : \mathcal{G} \to \mathbb{R}^+$ whenever expectations exist.
- **cumulative probability distributions:**

$$X \leq_{st} Y \Leftrightarrow F_X(a) \geq F_Y(a),\ \forall a \in \mathcal{G}.$$

- **probability mass functions:**

$$X \leq_{st} Y \Leftrightarrow \forall i,\ 1 \leq i \leq n,\ \sum_{k=i}^{n} d2(k) \leq \sum_{k=i}^{n} d1(k) \tag{1}$$

Notice that we use interchangeably $X \leq_{st} Y$ and $d2 \leq_{st} d1$.

Property 1. If $X \leq_{st} Y$, then for any increasing function f,

$$f(X) \leq_{st} f(Y)$$

The $\leq_{st}$ ordering is closed under mixture (Theorem 1.2.15 in p. 6 of [13]):

Theorem 1. *If X, Y and Θ are random variables such that $[X \mid \Theta = \theta] \leq_{st} [Y \mid \Theta = \theta]$ for all θ in the support of Θ, then $X \leq_{st} Y$.*

The following definition is used to compare Markov chains.

Definition 2. *Let $\{X(n),\ n \geq 0\}$ (resp. $\{Y(n),\ n \geq 0\}$) be a DTMC. We say $\{X(n),\ n \geq 0\} \leq_{st} \{Y(n),\ n \geq 0\}$, if $X(n) \leq_{st} Y(n),\ \forall n \geq 0$.*

Let $\mathbf{P}$ and $\mathbf{Q}$ be the probability transition matrix of $\{X(n), n \geq 0\}$ and $\{Y(n), n \geq 0\}$ respectively. If the chains are ergodic, let $\boldsymbol{\pi}_{\mathbf{P}}$ and $\boldsymbol{\pi}_{\mathbf{Q}}$ denote the corresponding steady state distributions, then $\boldsymbol{\pi}_{\mathbf{P}} \leq_{st} \boldsymbol{\pi}_{\mathbf{Q}}$. The following theorem provides sufficient conditions to establish the comparison of DTMCs.

Theorem 2. *Let $\mathbf{P}$ (resp. $\mathbf{Q}$) be the probability transition matrix of the time-homogeneous Markov chain $\{X(n), n \geq 0\}$ (resp. $\{Y(n), n \geq 0\}$). The comparison of Markov chains is established $\{X(n), n \geq 0\} \leq_{st} \{Y(n), n \geq 0\}$, if the following conditions are satisfied*

- $X(0) \leq_{st} Y(0)$,
- *at least one of the probability transition matrices is monotone, that is, either $\mathbf{P}$ or $\mathbf{Q}$ (say $\mathbf{P}$) is $\leq_{st}$ monotone, if for all probability vectors $\mathbf{p}$ and $\mathbf{q}$,*

$$\mathbf{p} \leq_{st} \mathbf{q} \Longrightarrow \mathbf{p}\mathbf{P} \leq_{st} \mathbf{q}\mathbf{P}$$

  which is equivalent to

$$1 \leq i \leq n-1, \quad \mathbf{P}[i, *] \leq_{st} \mathbf{P}[i+1, *]$$

  where $\mathbf{P}[i, *]$ denotes the row of matrix $\mathbf{P}$ for state i.
- *the transition matrices are comparable in the sense of the $\leq_{st}$ order:*

$$\mathbf{P} \leq_{st} \mathbf{Q} \Leftrightarrow 1 \leq i \leq n, \quad \mathbf{P}[i, *] \leq_{st} \mathbf{Q}[i, *]$$

3   Bounding Performance Measures Under Stationary Traffic

We present in this section the method we have developed in various publications [1–3].

3.1   Queue Model and Evolution Equations

Let us begin with some notation. The number of transmission units produced by the traffic source during the kth slot is denoted by $A(k)$, and $Q(k)$ and $D(k)$ are respectively the buffer length and the output (departure) traffic (flow) during the kth slot. The buffer size is noted by B and the service capacity during a slot by S. The input parameter $A(k)$ is specified by a discrete distribution (histogram), and the output parameters are also derived as histograms (Fig. 1).

The admission per packet is done with Tail Drop policy. Thus an arrival packet is accepted if there is a place in the buffer, otherwise it is rejected. The timing of events during a slot is as follows: arrivals occur first and they are followed immediately by services. The evolution equations for the buffer length $(Q(k))$ and the departure traffic $(D(k))$ can be given as follows:

$$Q(k) = \min(B, (Q(k-1) + A(k) - S)^+), \quad k \geq 1, \tag{2}$$

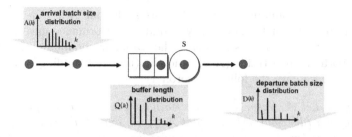

Fig. 1. Input and output parameters of a queueing model

where operator $(X)^+ = \mathtt{max}(X, 0)$.

$$D(k) = \mathtt{min}(S, Q(k-1) + A(k)), \quad k \geq 1. \tag{3}$$

The model of the queue is a time-inhomogeneous Discrete Time Markov Chains (DTMC), if the input arrivals are independent of the current queue state and the past of the arrival process. Under the stationary arrival assumptions, the underlying DTMC is time-homogenous.

The monotonicity of these equations under the $\leq_{st}$ order has been proved in [1,2]. Intuitively speaking, the monotonicity property states that if we consider two models under different arrival processes but comparable in the sense of the $\leq_{st}$ order, then the corresponding output parameters are also comparable in the sense of the $\leq_{st}$ order.

Let consider two queues. The first one is under arrival process $A(k)$, $k \geq 0$, and the output parameters (queue length, and departure traffic) noted by $Q(k)$ and $D(k), k \geq 0$. The second one is under arrival process $\tilde{A}(k)$, with output parameters: $\tilde{Q}(k)$, $\tilde{D}(k)$). At the beginning, $Q(0) \leq_{st} \tilde{Q}(0)$ and $D(0) \leq_{st} \tilde{D}(0)$. Without loss of generality, we assume that the queues are idle at $k = 0$, thus the queue lengths and the departure processes are empty, thus $Q(0) =_{st} \tilde{Q}(0)$ and $D(0) =_{st} \tilde{D}(0)$.

Theorem 3. *If* $A(k) \leq_{st} \tilde{A}(k)$, $\forall k > 0$, *then*

$$Q(k) \leq_{st} \tilde{Q}(k), \quad and \quad D(k) \leq_{st} \tilde{D}(k), \ \forall k > 0.$$

The monotonicity results follow from the fact that the $\leq_{st}$ order is associated to increasing functions and the underlying measures are defined by increasing functions of input parameters.

This theorem lets us to construct bounding systems. For instance, for a given system, let say the one under the arrival process $A(k)$, it is possible to construct bounding performance measures, $\tilde{Q}(k), \tilde{D}(k)$ by considering the bounding arrival process $\tilde{A}(k)$. Obviously, this approach is meaningful if the analysis under arrival $\tilde{A}(k)$ is more efficient to do. Notice that these are transient bounds thus the comparisons are satisfied at each instant k, and also for the steady state if it exists.

If a stationary bounding process $\tilde{\mathcal{A}}$ exists such that $A(k) \leq_{st} \tilde{A}$, $\forall k > 0$, it has been proved that the stationary bounding performance measures can be derived by considering the system under the stationary bounding process $\tilde{\mathcal{A}}$ [2]. Clearly if both the real traffic $(A(k))$ and the (upper) bounding traffic $(\tilde{A}(k))$, are stationary, we have the following corollary:

Corollary 1. *Let $\mathcal{A}$ (resp. $\tilde{\mathcal{A}}$) be the stationary exact (resp. upper bounding) input histogram (distribution) such that $\mathcal{A} \leq_{st} \tilde{\mathcal{A}}$, and $\mathcal{Q}, \mathcal{D}$ (resp. $\tilde{\mathcal{Q}}, \tilde{\mathcal{D}}$) be the stationary buffer length, departure flow under the exact $\mathcal{A}$, (resp. upper bounding $\tilde{\mathcal{A}}$) input arrival. If $Q(0) \leq_{st} \tilde{Q}(0)$, and $D(0) \leq_{st} \tilde{D}(0)$, then we have:*

$$\mathcal{Q} \leq_{st} \tilde{\mathcal{Q}} \ and \ \mathcal{D} \leq_{st} \tilde{\mathcal{D}}.$$

The lower bounding case can be similarly derived.

3.2   Bounding Histogram Construction

The complexity of the numerical analysis of performance measures (Eqs. 2 and 3) depends on the arrival distributions whatever the used method is. We advocate that, as the queue we model is stochastically monotone, it is possible to aggregate the input distribution (to reduce the number of atoms) for deriving in an easier way stochastic bounds on the performance measures. For a discrete distribution of probability, the complexity parameter is the number of atoms. Therefore we propose to apply the bounding approach to make the number of atoms smaller. The main advantage of this approach is the computation of bounds rather than approximations. Unlike approximations, the bounds allow us to have guarantees and check if QoS are satisfied or not.

Let the input arrival process is specified by a probability mass function (discrete distribution) $\boldsymbol{d}$ defined on N atoms. In [3], we have proposed an algorithm to build an upper and a lower bounding distribution, $\boldsymbol{d}1$ and $\boldsymbol{d}2$ with $n << N$ atoms. Moreover, $\boldsymbol{d}1$ and $\boldsymbol{d}2$ are the optimal bounds with respect to a given positive, increasing reward function, $\boldsymbol{r}$. Formally, for a given distribution $\boldsymbol{d}$ defined on $\mathcal{H}$ $(|\mathcal{H}| = N)$, we compute bounding distributions $\boldsymbol{d}1$ and $\boldsymbol{d}2$ defined respectively on $\mathcal{H}^u$, $\mathcal{H}^l$ $(|\mathcal{H}^u| = n, |\mathcal{H}^l| = n)$ such that:

1. $\boldsymbol{d}2 \leq_{st} \boldsymbol{d} \leq_{st} \boldsymbol{d}1$,
2. $\sum_{i \in \mathcal{H}} \boldsymbol{r}(i)\boldsymbol{d}(i) - \sum_{i \in \mathcal{H}^l} \boldsymbol{r}(i)\boldsymbol{d}2(i)$ is minimal among the set of distributions on n atoms that are stochastically lower than $\boldsymbol{d}$,
3. $\sum_{i \in \mathcal{H}^u} \boldsymbol{r}(i)\boldsymbol{d}1(i) - \sum_{i \in \mathcal{H}} \boldsymbol{r}(i)\boldsymbol{d}(i)$ is minimal among the set of distributions on n atoms that are stochastically upper than $\boldsymbol{d}$.

Notice that $\forall\, i \in \mathcal{H}$ and $i \notin \mathcal{H}^u$ (resp. $\forall\, i \in \mathcal{H}$ and $i \notin \mathcal{H}^l$), $\boldsymbol{d1}(i) = 0$ (resp. $\boldsymbol{d2}(i) = 0$) to establish the stochastic comparisons. Thus $\boldsymbol{d}1$ and $\boldsymbol{d}2$ denote the optimal bounding distributions on n atoms with respect to reward $\boldsymbol{r}$.

The proposed algorithm is based on dynamic programming and has a complexity of $O(N^2 n)$. Some heuristics with a smaller complexity which let to construct stochastic bounds with the required number of atoms but which are not in general optimal can be found in the same reference.

The number of atoms provide a trade-off between the accuracy of the bounds and the computation time. It can be determined in an incremental manner: one begins with a reduced number of atoms, if the accuracy of bounds is not satisfactory, the number of atoms can be incremented. The iteration can be stopped, if the required accuracy is reached and/or the computation time of bounds exceeds a fixed threshold.

Example. Let $d = [0.1, 0.4, 0.05, 0.15, 0.1, 0.2]$ be a discrete distribution defined on a support $\mathcal{H} = \{1, 2, 3, 4, 5, 6\}$ ($N = 6$). For reward function r sets to $r(i) = a_i, \forall\, a_i \in \mathcal{H}$, the expected reward of d is $R[d] = \sum_{a_i \in \mathcal{H}} r(i)\, d(i) = 3.35$.

The computation of the optimal stochastic upper bound $d1$ (resp. lower bound $d2$) of d with only 3 states (atoms) consists in exploring all 3 single hops paths from the largest (resp. smallest) atom and select the path for which $R[d1] - R[d]$ $(R[d] - R[d2])$ is the minimal.

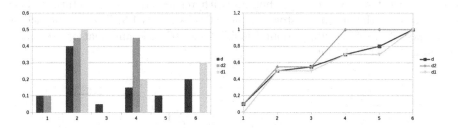

Fig. 2. Probability mass (left) and cumulative distribution functions (right).

We illustrate in Fig. 2, the probability mass functions and the cumulative distribution functions of the exact and the computed optimal bounding distributions ($d2 \leq_{st} d \leq_{st} d1$). The expected reward of the bounding distributions are: $R[d2] = 3.1$ and $R[d1] = 3.8$.

3.3   Performance Measure Bounds Under Stationary Arrivals

We are indeed interested in the performance analysis of the queue under real traffic traces. We present here an example given in [2] under stationary traffic assumption of real traces. In Fig. 3, a real traffic trace extracted from the MAWI traffic traces [16] is illustrated. Precisely, it corresponds to an IP traffic trace during one hour for a 150 Mbps transpacific line (samplepoint-F) for the 9th of January 2007 between 12:00 and 13:00. This traffic trace has an average rate of 109 Mbps. Using a sampling interval of T = 40 ms (25 samples per second), the resulting traffic trace has 90,000 frames (periods), an average of 4.37 Mb per frame and 80511 distinct values (atoms).

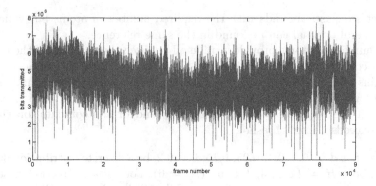

Fig. 3. MAWI traffic trace.

We present in Fig. 4, the lower and upper bounding histograms with $n = 10$ atoms for this trace, and the exact histogram without size reduction. The reward considered in the histogram size reduction algorithm is the identity function in order to construct optimal $\leq_{st}$ bounds with respect to the expectation. The expectation of the original histogram (noted as exact) is 4.3757×10^6 bits while the expectation of the upper bound is 4.5843×10^6 bits, and that of the lower bound is 4.1644×10^6 bits.

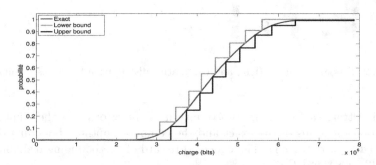

Fig. 4. Cumulative probability distributions (cdf) for the MAWI traffic.

In Fig. 5, the bounds on the blocking probability and the mean buffer length under MAWI traffic versus different values of reduction (atoms varying from 10 to 200) are given. In each figure, we give the results computed under: (1) the exact MAWI histogram (without reduction 80511 atoms), (2) lower bounding histogram and (3) upper bounding histogram. It can be seen that the bounds become tighter when the bounding histogram size increases. Thus a tradeoff between the accuracy of results and computation complexity can be found.

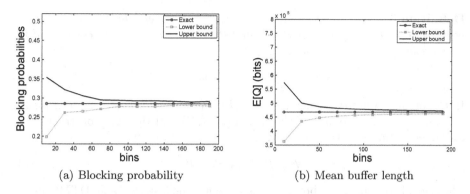

(a) Blocking probability   (b) Mean buffer length

Fig. 5. Accuracy versus the number of atoms: QoS parameters using the MAWI traffic

4   SBBP Input Traffic

We now consider that the traffic is modelled by a *Switched Batch Bernoulli Process* (SBBP) [7], and we show that our method can also be applied in this case. The SBBP process is an arrival process modulated by a Markov chain. This model is useful to characterise phase-dependent arrivals, i.e. the arrival processes during different phases have different characteristics. If there are p arrival phases, the phase evolution is controlled by a time-homogeneous DTMC defined on state space $\mathcal{P} = \{1, \ldots, p\}$. Let $\mathbf{F}$ be the probability transition matrix for phase changes, then $\mathbf{F}(i, j)$ is the probability of the transition from phase i to phase j.

The state of the system at time k can be denoted by $QP(k) = (Q(k), \phi(k))$. The first component $Q(k)$ is the number of entities in the buffer and $\phi(k)$ is the arrival phase during slot k. In each arrival phase $i \in \mathcal{P}$, the arrival process $\mathcal{A}^i$ is assumed to be stationary and independently, identically distributed. The underlying system $\{QP(k), k \geq 0\}$ is a time-homogeneous DTMC. During time k, the arrival phase is $\phi(k)$, and the evolution of $Q(k)$ is the same as in the stationary arrival case, but under arrival $\mathcal{A}^{\phi(k)}$ instead of $\mathcal{A}$. Thus, $Q(k)$ takes values in the set $\mathcal{N} = \{0 \cdots B\}$, and evolves as follows:

$$Q(k+1) = \mathtt{min}\left(B, (Q(k) + \mathcal{A}^{\phi(k)} - S)^+\right).$$

The evolution of the second component, $\phi(k)$ is controlled by a Markov chain. The state space of $\{QP(k), k \geq 0\}$ is the product space $\mathcal{S} = \mathcal{N} \times \mathcal{P}$.

4.1   Bounds Under SBBP Input Traffic

We construct the bounding models by fixing the arrival phase, and the comparisons are established arrival phase by arrival phase. The partial order $\preceq$ on $\mathcal{S}$ to compare any two states $x, y \in \mathcal{S}$ is defined as following:

Definition 3. *Let* $x = (x_q, x_\phi), y = (y_q, y_\phi) \in \mathcal{S}$, *where the first components correspond to the buffer lengths (Q) and the second components correspond to the arrival phases (ϕ).*

$$x \preceq y \quad \text{iff} \quad x_q \leq y_q \quad \text{and} \quad x_\phi = y_\phi.$$

In the bounding system denoted by $\tilde{Q}P(k)$, the arrival processes in each phase ($\tilde{\mathcal{A}}^i$) are the upper bounds of the real traffic ($\mathcal{A}^i$) and they are constructed as explained in Subsect. 3.2. Formally,

$$\forall i \in \mathcal{P}, \quad \mathcal{A}^i \leq_{st} \tilde{\mathcal{A}}^i. \tag{4}$$

The second component of both models are controlled by the same DTMC independently of the first component. We assume that at the beginning, $k = 0$,

$$(Q(0), \phi(0)) =_{st} (\tilde{Q}(0), \tilde{\phi}(0)).$$

Thus, if we start with the same initial states in both models, the evolution of the second component will be the same at each time k.

Corollary 2. *Let* $\mathcal{Q}, \mathcal{D}$ *be the steady-state marginal distributions of the buffer length and the departure flow under arrival distributions* $\mathcal{A}^i$, *while* $\tilde{\mathcal{Q}}$ *and* $\tilde{\mathcal{D}}$ *denote the corresponding distributions under the upper bounding arrival distributions* $\tilde{\mathcal{A}}^i$.
*   If $\mathcal{A}^i \leq_{st} \tilde{\mathcal{A}}^i, \quad \forall i \in \mathcal{P}$, then*

$$\mathcal{Q} \leq_{st} \tilde{\mathcal{Q}} \quad \text{and} \quad \mathcal{D} \leq_{st} \tilde{\mathcal{D}}.$$

Proof. By fixing the arrival phase, we derive bounds on conditional distributions. At each time k, for all arrival phases $i \in \mathcal{P}$, we have:

$$[Q(k) \mid \phi = i] \leq_{st} [\tilde{Q}(k) \mid \phi = i] \text{ and } [D(k) \mid \phi = i] \leq_{st} [\tilde{D}(k) \mid \phi = i].$$

As the $\leq_{st}$ ordering is closed under mixtures (Theorem 1 in Sect. 2), we have the comparison of the marginal distributions at each time k:

$$Q(k) \leq_{st} \tilde{Q}(k) \text{ and } D(k) \leq_{st} \tilde{D}(k).$$

By construction, the steady-states exist, then it follows from the convergence in distribution:

$$\mathcal{Q} \leq_{st} \tilde{\mathcal{Q}} \quad \text{and} \quad \mathcal{D} \leq_{st} \tilde{\mathcal{D}}.$$

4.2 Numerical Analysis

Due to the SBBP arrivals, we have a block structured Markov chain.

$$\mathbf{P} = \begin{pmatrix} P_{11} & P_{12} & \cdots & P_{1p} \\ P_{21} & P_{22} & \cdots & P_{2p} \\ \vdots & \vdots & \ddots & \vdots \\ P_{p1} & P_{p2} & \cdots & P_{pp} \end{pmatrix}.$$

Let (x_q, x_q) and (y_q, y_p) be two states of the DTMC, and $\mathbf{R}_\phi$ be the transition matrix of the system when the arrivals are in phase ϕ. The transition matrix $\mathbf{P}$ of the Markov chain $(QP(k))$ is

$$\mathbf{P}((x_q, x_p), (y_q, y_p)) = \mathbf{F}(x_p, y_p)\mathbf{R}_{x_p}(x_q, y_q).$$

where $\mathbf{F}$ is the transition matrix of phase modulation.

Such a structured matrix is denoted as a functional Kronecker product in the theory of Stochastic Automata Networks [6,17]. It has many important properties which can be taken into account to obtain efficient numerical techniques.

Property 2. The Markov chain $(QP(k))$ of the model with SBBP arrival is lumpable according to the partition defined by the phase of the arrival process.

Let $\pi_\mathbf{P}$ (resp. $\pi_\mathbf{F}$) be the steady-state distribution for matrix $\mathbf{P}$ (resp. $\mathbf{F}$). We know that the lumpability implies that there exists p vectors ψ_j of size $B + 1$, denoting the conditional queue length probabilities when the arrival phase is j. The stationary distribution $\pi_\mathbf{P}$ is then computed as follows:

$$\pi_\mathbf{P}(i) = \sum_{j=1}^{p} \pi_\mathbf{F}(j)\,\psi_j(i), \quad \forall\, i = 0\cdots B.$$

To compute the steady-state solution of the model, we use the Iterative Aggregation Disaggregation (IAD) algorithm specialised for lumpable matrices published in [6] to obtain successive values of vectors ψ_i which are denoted $\psi_i^{(t)}$ at iteration t. This algorithm is based on the following steps.

1. Initialise $\psi_i^{(0)}$, for all i
2. Compute $\pi_\mathbf{F}$, the steady state probability vector of $\mathbf{F}$
3. Compute vectors $\psi_i^{(t+1)}$ using a Block Gauss Seidel iteration for matrix $\mathbf{P}$ in block form:
   (a) $Z_i^{(t+1)} = \pi_\mathbf{F}(i)\dfrac{\psi_i^{(t)}}{||\psi_i^{(t)}||_1}, \; \forall\, i = 1\cdots p$
   (b) $\psi_i^{(t+1)} = \psi_i^{(t)}\mathbf{P}_{ii} + \sum_{j=i+1}^{p} Z_j^{(t+1)}\mathbf{P}_{ji} + \sum_{j=1}^{i-1} \psi_j^{(t+1)}\mathbf{P}_{ji}, \; \forall\, i = 1\cdots p$
4. Normalise vectors $\psi_i^{(t+1)}$ to be distributions of probability
5. If $\sum_i ||\psi_i^{(t+1)} - \psi_i^{(t)}||_\infty$ is smaller than a threshold, go to step 6. Otherwise set $t = t + 1$ and go to step 3.
6. Compute $\pi_\mathbf{P}$, such that: $\pi_\mathbf{P}(i) = \sum_{j=1}^{p} \pi_\mathbf{F}(j)\,\psi_j^{(t)}(i), \quad \forall\, i = 0\cdots B.$

Theoretically, in the first step we can initialise vectors $\psi_i^{(0)}$ with any distribution of probability. Taking into account the properties of the arrivals during the phases as defined in the next paragraph, we have used three phases and the following guess: $\psi_1^{(0)} = \delta_0$, $\psi_3^{(0)} = \delta_B$, and $\psi_2^{(0)}$ equal to the steady state probability vector of matrix $\mathbf{R}_2$ (transition matrix of the system when the arrivals are in phase 2).

4.3 Numerical Results

In order to illustrate the results stated in this paper, we propose to compute the performance measures of a finite single queue under real traffic trace modelled as SBBP arrival process and constant service. We consider the MAWI trace [16] which corresponds to a little more than 10 h of an IP traffic on transpacific line with link capacities of 128 Kbps, carried between the 6th of march 2007 at 18 : 00 and the 7th of march 2007 at 4 : 24 : 27. For a sampling period $T = 40$ ms, we obtain the trace shown in Fig. 6 with 922873 frames and 4579 different atoms.

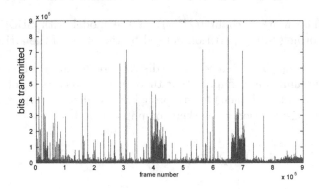

Fig. 6. MAWI traffic trace (more than 10 h).

We distinguish three phases, phase 1 corresponds to low traffic, phase 2 to medium traffic, and phase 3 to heavy traffic. We assume that for each slot, the traffic trace is characterised by its volume per sampling period. If the traffic per sampling period is less or equal to the minimum threshold (10 Kbps), the arrival phase is 1, and if it is greater or equal to the maximum threshold (100 Kbps), the arrival phase is 3. When the traffic is between the thresholds, the arrival phase is 2. In each phase, the traffic is defined by a stationary arrival process associated to this phase. The probability transition matrix for phase modulation is defined as follows:

$$\mathbf{F}(i, j) = \frac{\text{number of transition between phase } i \text{ and phase } j}{\text{number of slots in phase } i}.$$

The resulting transition matrix of phases $\mathbf{F} = \begin{pmatrix} 0.9982 & 0.0018 & 0.0000 \\ 0.5563 & 0.4163 & 0.0274 \\ 0.2706 & 0.2615 & 0.4679 \end{pmatrix}$.

The histogram of each phase is defined respectively on 1228 atoms (phase 1), 2568 atoms (phase 2) and 783 atoms (phase 3). They are characterised by the following statistical descriptions (Table 1):

Let us emphasise here that our goal is not to study how to obtain an accurate SBBP model for a given trace. We just aim to construct such a model to apply our

Table 1. Statistical descriptions of the considered MAWI traffic trace

	Expected value (bits)	Standard deviation (bits)	Coefficient of variation
Phase 1	433.56	1.0503×10^3	5.8684
Phase 2	28953	2.13×10^4	0.5413
Phase 3	2.1515×10^5	1.2844×10^5	0.3564

bounding algorithms and explain how our approach works and can be accurate if the input arrival is a SBBP process. The thresholds have been arbitrarily chosen. The statistical analysis of traces to derive fitting models is out of the scope of this paper.

We now apply our numerical bounding approach to this model to obtain two performance measures (expected buffer length and blocking probability) versus the buffer size (B) which varies from 100 Kb to 3 Mb. We consider a deterministic service capacity of 35 Kbps. The bounding histograms (noted by $L.b$ for the lower bound and by $U.b$ for the upper bound) are constructed on reduced state space with 100 atoms. The exact results (without reduction) and the bounds of these performance measures are given in Table 2. The results by assuming stationary traffic are presented in Table 3. The computation times in seconds are given in Table 4.

Table 2. Blocking probabilities and expected buffer lengths versus buffer size

B	Blocking probabilities			Expected buffer length		
	Exact	L.b	U.b	Exact	L.b	U.b
10^5	0.0032218	0.0031552	0.0035345	19297.7	17628.2	19448.6
2×10^5	0.0021574	0.0020811	0.0022456	46352.2	41679.7	46696.1
5×10^5	0.0012534	0.0011796	0.0013074	154693	137686	156254
10^6	0.0008447	0.0007416	0.0008902	401307	351574	405695
2×10^6	0.0005545	0.0004148	0.0005890	975858	813564	985124
5×10^6	0.0003562	0.0001691	0.0003835	3046090	2205710	3077930

We observe that the computed bounds under SBBP arrivals are relevant, especially upper bounds are quite tight for this example. Moreover the accuracy of bounds is not degraded when the histogram sizes increase. In terms of complexity, we remark that the computation times of bounds are significantly less than the exact one (the computation time is divided approximately by three when $B = 10^6$, by four when $B = 2 \times 10^6$, and by five for $B = 5 \times 10^6$). These results show that this approach provides an interesting tradeoff between the accuracy of results and the computational complexity to check if some QoS requirements are satisfied or not under SBBP arrivals.

Table 3. Performance measuresunder stationary traffic assumption (without reduction)

B	BP	E[Q]
10^5	0.00419	21651.7
2×10^5	0.00238	51641.8
5×10^5	0.00101	147084
10^6	0.000295	260630
2×10^6	1.75742e−05	304474
5×10^6	1.62373e−10	306020

Table 4. Computation times in second

B	SBBP input			Stationary input
	Exact	L.b	U.b	Exact
10^5	2.58	2.41	2.29	78.6
2×10^5	5.584	4.45	3.89	554.55
5×10^5	41.94	17.37	17.70	3710.7
10^6	203.61	74.08	79.57	7564.05
2×10^6	1180.62	359.61	422.9	13736.2
5×10^6	14085	3325	3695	44999.4

Regarding the difference between stationary input traffic and SBBP traffic, we note that the blocking probabilities and the expected buffer length are much greater under SBBP traffic especially for the large buffer sizes. This phenomenon is due to the higher variance for the SBBP arrival process.

5   Conclusion

The stochastic performance bounds of a queue under stationary histogram-based input traffics is generalised to the Markov modulated arrivals. The traffic is assumed to be stationary during a phase and the traffic phase transition is controlled by a DTMC. We illustrate the applicability of this approach by giving some numerical results for a system with arrivals derived from a real traffic trace. We want to emphasise that despite the bivariate process we can use strong stochastic bounds rather than weak or weak* comparisons (see [13]). The techniques we develop here and the associated publications [1,2] lead to an algorithmic analysis of queues based on measurements for the arrival process.

References

1. Aït-Salaht, F., Castel-Taleb, H., Fourneau, J.-M., Pekergin, N.: Stochastic bounds and histograms for network performance analysis. In: Balsamo, M.S., Knottenbelt, W.J., Marin, A. (eds.) EPEW 2013. LNCS, vol. 8168, pp. 13–27. Springer, Heidelberg (2013). doi:10.1007/978-3-642-40725-3_3

2. Aït-Salaht, F., Castel Taleb, H., Fourneau, J.-M., Pekergin, N.: Performance analysis of a queue by combining stochastic bounds, real traffic traces and histograms. Comput. J. **59**(12), 1817–1830 (2016)
3. Aït-Salaht, F., Cohen, J., Castel Taleb, H., Fourneau, J.M., Pekergin, N.: Accuracy vs. complexity: the stochastic bound approach. In: 11th International Workshop on Discrete Event Systems, pp. 343–348 (2012)
4. Fischer, W., Hellstern, K.M.: The Markov-modulated Poisson process (MMPP) cookbook. Perform. Eval. **18**, 149–171 (1992)
5. Gupta, V., Harchol-Balter, M., Dai, J.G., Zwart, B.: On the inapproximability of $M/G/K$: why two moments of job size distribution are not enough. Queueing Syst. **64**(1), 5–48 (2010)
6. Gusak, O., Dayar, T., Fourneau, J.-M.: Iterative disaggregation for a class of lumpable discrete-time stochastic automata networks. Perform. Eval. **53**(1), 43–69 (2003)
7. Hashida, O., Takahashi, Y., Shimogawa, S.: Switched batch Bernoulli process (SBBP) and the discrete-time SBBP/G/1 queue with application to statistical multiplexer performance. IEEE J. Select. Areas Commun. **9**(3), 394–401 (1991)
8. Hernández-Orallo, E., Vila-Carbó, J.: Network performance analysis based on histogram workload models. In: MASCOTS, pp. 209–216 (2007)
9. Hernández-Orallo, E., Vila-Carbó, J.: Web server performance analysis using histogram workload models. Comput. Netw. **53**(15), 2727–2739 (2009)
10. Hernández-Orallo, E., Vila-Carbó, J.: Network queue and loss analysis using histogram-based traffic models. Comput. Commun. **33**(2), 190–201 (2010)
11. Horváth, G., Telek, M., Buchholz, P.: A map fitting approach with independent approximation of the inter-arrival time distribution and the lag correlation. In: QEST, pp. 124–133. IEEE Computer Society (2005)
12. Klemm, A., Lindemann, C., Lohmann, M.: Traffic modelling of IP networks using the batch Markovian arrival process. Perform. Eval. **54**(25), 149–173 (2003)
13. Muller, A., Stoyan, D.: Comparison Methods for Stochastic Models and Risks. Wiley, New York (2002)
14. Muscarielloa, L., Melliaa, M., Meoa, M., Ajmone, M.M., Lo Cignob, R.: Markov models of internet traffic and a new hierarchical MMPP model. Comput. Commun. **28**, 1835–1851 (2005)
15. Skelly, P., Schwartz, M., Dixit, S.S.: A histogram-based model for video traffic behaviour in an ATM multiplexer. IEEE/ACM Trans. Netw. **1**(4), 446–459 (1993)
16. Sony, K.C., Cho, K.: Traffic data repository at the wide project. In: Proceedings of USENIX 2000 Annual Technical Conference: FREENIX Track, pp. 263–270 (2000)
17. Stewart, W.: Introduction to the numerical Solution of Markov Chains. Princeton University Press, New Jersey (1995)
18. Wittevrongel, S., Bruneel, H.: Discrete-time queues with correlated arrivals and constant service times. Comput. Oper. Res. **26**, 93–108 (1999)
19. Zhou, W., Wang, A.: Discrete-time queue with Bernoulli bursty source arrival and generally distributed service times. Appl. Math. Model. **3**, 2223–2240 (2013)

Equivalence and Lumpability of FSPNs

Falko Bause[1], Peter Buchholz[1(✉)], Igor V. Tarasyuk[2], and Miklós Telek[3]

[1] Informatik 4, TU Dortmund, 44221 Dortmund, Germany
`{falko.bause,peter.buchholz}@cs.tu-dortmund.de`
[2] A.P. Ershov Institute of Informatics Systems,
Siberian Branch of the Russian Academy of Sciences,
Acad. Lavrentiev pr. 6, 630090 Novosibirsk, Russian Federation
`itar@iis.nsk.su`
[3] MTA-BME Information Systems Research Group,
Department of Networked Systems and Services,
Technical University of Budapest, 1521 Budapest, Hungary
`telek@hit.hme.hu`

Abstract. We consider equivalence relations for Fluid Stochastic Petri Nets (FSPNs). Based on equivalence relations for Stochastic Petri Nets (SPNs), which are derived from lumpability for Markov Chains, and from lumpability for certain classes of differential equations, we define an equivalence relation for FSPNs. Lumpability for the differential equations is based on a finite discretization approach and permutations of the fluid part of the FSPN.

As for other modeling formalisms, the availability of an appropriate equivalence relation allows one to aggregate sets of equivalent states into single states. This state space reduction can be exploited for a more efficient analysis of FSPNs using a discretization approach. Lumpable equivalence relations can be computed from an appropriately discretized state space of the stochastic process or directly from the FSPN.

Keywords: Fluid Stochastic Petri Nets · Lumpability · Equivalence

1 Introduction

The idea of lumping states in a discrete system has a long history in Markov chains [1,15] but has also been used in linear systems [6] and for differential equations [16]. Later it has been applied to specific modeling formalisms like stochastic process algebras [11] and even stochastic Petri nets [2]. Current developments can be found for fluid models [13,17]. The central idea of lumpability is the definition of classes of states with an identical behavior and the substitution of the state classes by single states without altering the behavior of the system as it is observed. We present this approach for FSPNs in two versions. First we introduce a discretized version of the system and discuss lumping on the

---

The work of F. Bause, P. Buchholz and I. Tarasyuk has been partially supported by DFG under grant BE 1267/14-1.

N. Thomas and M. Forshaw (Eds.): ASMTA 2017, LNCS 10378, pp. 16–31, 2017.
DOI: 10.1007/978-3-319-61428-1_2

discretized model. Next we discuss lumping on the original FSPN. The first approach involves identical behavior on the level ODEs while the latter one presents identical behavior on the level of PDEs.

New Contribution of the Paper: Lumpability has been applied in the above mentioned papers for discrete models or for specific types of continuous models as they result from kinetic differential equations [16]. The latter approach has then been used as a basis to define lumpability for a fluid description of stochastic process algebra terms which result from a large number of identical and symmetric components [13,17]. Our approach combines lumpability for discrete and for continuous systems and presents, to the best of our knowledge for the first time, an approach that can be applied to hybrid systems. Additionally, the lumping of the continuous part goes beyond the approach presented for stochastic process algebras because lumping does not necessarily imply symmetry in the model.

The rest of the paper is organized as follows. FSPNs and the notation are introduced in Sect. 2. The analytical description of FSPNs by means of PDEs and a proposed discretizations approach resulting in an ODE-based analytical description are provided in Sect. 3. Lumpability of the discretized system is analyzed in Sect. 4, while in Sect. 5, lumpability is analyzed directly on the system matrices without the discretization step. The paper is concluded in Sect. 6.

2   Background and Definitions

We consider a class of FSPNs, which is similar to FSPNs, presented in [10,12]. A FSPN is an 7-tuple $(P, T, \bar{m}_0, A, B, F, R)$, where

- P is the set of places which is subdivided into the set of discrete places P_d and the set of continuous places P_c,
- T is the set of (timed) transitions,
- $\bar{m}_0 = (\boldsymbol{m}_0, \boldsymbol{x}_0)$ is the initial marking, where $\boldsymbol{m}_0 \in \mathbb{N}^{|P_d|}$ is a vector containing the number of tokens on each discrete place and $\boldsymbol{x}_0 \in \mathbb{R}^{|P_c|}$ is a vector which contains for each continuous place the level of fluid at the place. Let $\mathcal{M}_d$ be the set of all reachable discrete markings, $\mathcal{M}_c$ be the set of reachable continuous markings and $\mathcal{M}$ the set of all markings,
- A is the set of arcs which is subdivided into discrete arcs A_d : $((P_d \times T) \cup (T \times P_d)) \to \mathbb{N}$ (where A_d defines the multiplicity of the arc) and continuous arcs $A_c : (P_c \times T) \cup (T \times P_c) \to \{0, 1\}$,
- B is the set of capacities of fluid places, i.e., $B : P_c \to \mathbb{R}_{>0}$,
- F the set of transition rates which is a function $F : T \times \mathcal{M} \to \mathbb{R}_{\geq 0}$,
- R the set of flow rates which is a function $R : A_c \times \mathcal{M} \to \mathbb{R}_{\geq 0}$.

We do not consider immediate transitions here which are commonly available in FSPNs (e.g., [10,12]), because it is easier to define equivalence relations for FSPNs with only timed transitions. However, it is possible to extend the approach to FSPNs with immediate transitions. The marking dependent fluid rate is a very powerful concept. It allows one to model for example inhibitor arcs or place capacities for discrete places, both are not explicitly part of our class of nets.

For a transition $t \in T$, we denote the input places by $\bullet t = \{p \in P_d | A_d(p, t) > 0\}$, similarly the output places by $t\bullet = \{p \in P_d | A_d(t, p) > 0\}$. For continuous places the notation $\circ t = \{p \in P_c | A_c(p, t) = 1\}$ and $t\circ = \{p \in P_c | A_c(t, p) = 1\}$ are applied. The input and output transitions, $\bullet p, p\bullet, \circ p$ and $p\circ$, are defined similarly.

A transition $t \in T$ is enabled in marking $\bar{m} = (\boldsymbol{m}, \boldsymbol{x})$, if for all $p \in \bullet t$, $A_d(p, t) \leq \boldsymbol{m}(p)$ and $F(t, \bar{m}) > 0$. Let $ena(\bar{m})$ be the set of transitions enabled in marking $\bar{m}$. Enabled transitions may modify the discrete and continuous state (i.e., marking) of the net.

We start with the discrete part of the marking. The discrete part is modified by firing an enabled transition. Firing times are exponentially distributed with rate $F(t, \bar{m})$ for $t \in ena(\bar{m})$. The transition that fires is selected according to a race condition. Firing transition t in marking $\bar{m} = (\boldsymbol{m}, \boldsymbol{x})$ results in the new marking $\bar{m}' = (\boldsymbol{m}', \boldsymbol{x})$ with $\boldsymbol{m}'(p) = \boldsymbol{m}(p) - A_d(p, t) + A_d(t, p)$ for all $p \in P_d$. The enabling conditions of transitions assure that all components of m' are non-negative. We use the notation $\bar{m} \xrightarrow{t} \bar{m}'$ if t fires in $\bar{m}$ and results in marking $\bar{m}'$. If only the discrete part is relevant we use the notation $\boldsymbol{m} \xrightarrow{t} \boldsymbol{m}'$. Observe that the firing of transitions does not modify the continuous state.

The continuous marking, $\boldsymbol{x}$, is continuously modified with a finite rate by enabled transitions, as long as the place capacities are respected. In marking $\bar{m}$, the potential flow rate for place $p \in P_c$ is given by

$$\check{r}_p(\bar{m}) = \sum_{t \in ena(\bar{m}) \cap \circ p} R((t, p), \bar{m}) - \sum_{t \in ena(\bar{m}) \cap p\circ} R((p, t), \bar{m}). \quad (1)$$

The actual flow rate has to take care of the place capacities and is defined as

$$r_p(\bar{m}) = \begin{cases} \check{r}_p(\bar{m}) & \text{if } 0 < x_p < B(p), \\ \check{r}_p(\bar{m}) & \text{if } x_p = 0 \wedge \check{r}_p(\bar{m}) > 0, \\ \check{r}_p(\bar{m}) & \text{if } x_p = B(p) \wedge \check{r}_p(\bar{m}) < 0, \\ 0 & \text{otherwise,} \end{cases} \quad (2)$$

where x_p is the fluid level at fluid place p. The rate describes the flow rate into a continuous place, i.e., $r_p(\bar{m}) = \frac{dx_p(\tau)}{d\tau}$, where τ denotes the time, and negative flow rate represents a decaying fluid level.

The model allows one to define some nasty behaviors, which means that flows or rates change infinitely often in a finite interval, as for example shown in [5]. We will exclude these behaviors in the following section and assume that the majority of the transition rate and flow rate functions are either independent or a piecewise constant function of the continuous marking $\boldsymbol{x}$. In principle, simulation can be applied to analyze FSPNs. We consider here numerical analysis via discretization where lumping helps to reduce the analysis complexity. To describe our approach we introduce several restrictions for the allowed class of nets. Some of these restrictions may be relaxed and still allow one to compute equivalence relations and analyze the resulting systems numerically, others are essential in the sense that otherwise a numerical analysis is no longer possible and an equivalence relation to reduce the state space can no longer be computed. The approach will be presented in the subsequent sections.

3 Discretization and Analysis

We consider only FSPNs with a finite set $\mathcal{M}_d$ otherwise numerical analysis can only be applied in very specific cases. Generation of the set $\mathcal{M}_d$ is in general non-trivial due to the presence of marking dependent transition rates which may become zero. However, it is easy to compute a super-set of $\mathcal{M}_d$ by neglecting all continuous components in the net which means that enabling conditions of transitions that depend on the filling of fluid places are simply ignored. We assume in the sequel that $\mathcal{M}_d$ or an appropriate finite super-set of $\mathcal{M}_d$ can be generated using common algorithms for state space generation.

For $\boldsymbol{x} \in \mathcal{M}_c$, $\boldsymbol{Q}(\boldsymbol{x})$ is a $|\mathcal{M}_d| \times |\mathcal{M}_d|$ matrix including the transitions rates if the continuous marking is $\boldsymbol{x}$. Markings from $\mathcal{M}_d$ are numbered consecutively from 1 through $|\mathcal{M}_d|$. We use the marking $\boldsymbol{m}_i$ and its number i interchangeably and have for the elements of matrix $\boldsymbol{Q}(\boldsymbol{x})$

$$q_{ij}(\boldsymbol{x}) = \sum_{t \in ena(\boldsymbol{m}_i) \wedge \boldsymbol{m}_i \xrightarrow{t} \boldsymbol{m}_j} F(t, (\boldsymbol{m}_i, \boldsymbol{x})), \text{ for } i \neq j,$$
$$q_{ii}(\boldsymbol{x}) = -\sum_{j \neq i} q_{ij}(\boldsymbol{x}). \tag{3}$$

Similarly, we define for each continuous place $p \in P_c$ a diagonal matrix $\boldsymbol{R}_p(\boldsymbol{x})$ of size $|\mathcal{M}_d| \times |\mathcal{M}_d|$ with $r_p((\boldsymbol{m}_i, \boldsymbol{x}))$ in position (i, i).

FSPNs as we defined them allow for a very complex behavior where the flow rate and also transition rates depend on the filling of fluid places in an arbitrary complex way. In full generality, such a behavior can hardly be analyzed. Therefore we assume that $\mathcal{M}_c$ can be decomposed in finitely many disjoint subsets $\mathcal{M}_c^1, \ldots, \mathcal{M}_c^K$ such that for $\boldsymbol{x}, \boldsymbol{y} \in \mathcal{M}_c^k$ $\boldsymbol{Q}(\boldsymbol{x}) = \boldsymbol{Q}(\boldsymbol{y})$ and $\boldsymbol{R}_p(\boldsymbol{x}) = \boldsymbol{R}_p(\boldsymbol{y})$ for all $p \in P_c$ and $\boldsymbol{m} \in \mathcal{M}_d$. We assume that each subset $\mathcal{M}_c^k$ is built from finite intervals (b_p^{k-1}, b_p^k) with $b_p^{k-1} < b_p^k$ $(1 \leq k < K)$, $b_p^0 = 0, b_p^K = B(p)$ for $p \in P_c$. We note that probability mass of various dimensions, characterized by appropriate boundary equations, can develop at set boundaries, if some components of $\boldsymbol{R}_p(\boldsymbol{x})$ changes sign, but their discussion we also neglect here. Thus, we assume that at b_p^k the functions in the matrices $\boldsymbol{Q}(\boldsymbol{x})$ and $\boldsymbol{R}_p(\boldsymbol{x})$ are left or right continuous or appropriate boundary conditions can be defined.

The dynamic behavior of FSPNs with more than one continuous place specifies a set of partial differential equations. The derivation for these equations will be briefly summarized and follows [3,8–10,12]. The transient behavior starting from $\bar{m}_0$ is considered. Let $M(\tau)$, $X(\tau)$ be the processes describing the evaluation of the discrete and continuous marking, respectively. The following notations are used for $\boldsymbol{m}_i \in \mathcal{M}_d$, $\boldsymbol{x} \in \mathcal{M}_c$ and time $\tau \geq 0$:

– $\pi_i(\tau) = Prob\,(M(\tau) = \boldsymbol{m}_i)$ are the discrete state probabilities,
– $H_i(\tau, \boldsymbol{x})$ is the CDF of the fluid level at fluid places when the discrete state is $\boldsymbol{m}_i$, we have

$$\pi_i(\tau) = \int_0^{B(p_1)} \cdots \int_0^{B(p_{|P_c|})} H_i(\tau, (dx_1, \ldots, dx_{|P_c|})),$$

– $f(\tau, \boldsymbol{x}) = \sum_{p_i \in P_d} h_i(\tau, \boldsymbol{x})$ the fluid density.

For an x where $H_i(\tau, x)$ is continuous $h_i(\tau, x) = \frac{\partial}{\partial x_1} \cdots \frac{\partial}{\partial x_{|P_c|}} H_i(\tau, x)$ is the probability density of the fluid places. The densities $h_i(\tau, x)$ are collected in a vector $h(\tau, x)$ of length $|M_d|$. The dynamic behavior of the system is described by the following set of partial differential equation [12, Theorem 1]

$$\frac{\partial h(\tau, x)}{\partial \tau} + \sum_{p \in P_c} \frac{\partial (h(\tau, x) R_p(x))}{\partial x_p} = h(\tau, x) Q(x). \tag{4}$$

For an x where $H_i(\tau, x)$ is not continuous probability mass develops in various dimensions. These probability masses (e.g., when $p_1 \in P_c$ is at its lower boundary, $p_2 \in P_c$ is at its upper boundary and $p_3 \in P_c$ is between its boundaries), whose number is exponentially increasing with the number of fluid places, are characterized by the boundary equations. Here we avoid the discussion of those boundary equations by referring to [12, Theorem 1], where multi-dimensional masses are considered at the lower boundaries of fluid places.

Results are computed in terms of discrete and continuous markings. We define two functions $g_d : M_d \to \mathbb{R}_{\geq 0}$ and $g_c : M_c \to \mathbb{R}_{\geq 0}$ that indicate the gain or reward with respect to the discrete or continuous state. We assume that the intervals M_c^k are defined such that for $x, y \in M_c^k$ $g_c(x) = g_c(y)$. The expectation of the overall gain at time τ, $G(\tau)$, is then given by

$$E(G(\tau)) = \sum_{m_i \in M_d} \left(\pi_i(\tau) g_d(m_i) + \int_{x_1} \cdots \int_{x_{|P_c|}} g_c(x) H_i(\tau, (dx_1, \ldots, dx_{|P_c|})) \right).$$

To analyze the system numerically, a discretization approach is applied. We introduce a simple first-order scheme following the ideas presented in [3,8,12]. Let Δ_p the discretization step for place $p \in P_c$. We assume that $B(p)$ is a multiple of Δ_p and $n_p = B(p)/\Delta_p$. Let $\Delta = (\Delta_1, \ldots, \Delta_{|P_c|})$ be a discretization scheme. Discretization defines a finite state space S_Δ with $n_\Delta = |M_d| \prod_{p \in P_c} n_p$ states. Each state is defined by a vector $(u_0, u_1, \ldots, u_{|P_c|})$ of length $1 + |P_c|$ where $u_0 \in M_d$ and $u_p \in \{1, \ldots, n_p\}$ for $p \in P_c$. States in S_Δ are ordered lexicographically according to their vector representation. Depending on the context, we use the vector representation for states or their number in the state space.

To compute transition rates in the discretized state space different methods exist. We apply a finite volume method and start with the transition rates of the discrete part, as follows

$$q_{ij}^k \triangleq q_{ij}^{(k_1, \ldots, k_{|P_c|})} = \frac{1}{\prod_{p \in P_c} \Delta_p} \int_{(k_1-1)\Delta_1}^{k_1 \Delta_1} \cdots \int_{(k_{|P_c|}-1)\Delta_{|P_c|}}^{k_{|P_c|}\Delta_{|P_c|}} q_{ij}(x) dx_1, \ldots dx_{|P_c|}, \tag{5}$$

where $k = (k_1, \ldots, k_{|P_c|})$ is ranging from $(1, \ldots, 1)$ to $(n_1, \ldots, n_{|P_c|})$. Since the function $q_{ij}(x)$ is piecewise constant, the integral can be evaluated as a sum. Matrix

$$\bar{Q} = \begin{pmatrix} \bar{Q}_{1,1} & \cdots & \bar{Q}_{1,|M_d|} \\ \vdots & \ddots & \vdots \\ \bar{Q}_{|M_d|,1} & \cdots & \bar{Q}_{|M_d|,|M_d|} \end{pmatrix} \tag{6}$$

is a $n_\Delta \times n_\Delta$ generator matrix of a Markov chain, where $\bar{Q}_{ij} = diag\left(q_{ij}^k\right)$ is a diagonal matrix.

For the discretized flow rates define

$$r_{i,k}^p = \frac{1}{\prod_{p \in P_c} \Delta_p} \int_{(k_1-1)\Delta_1}^{k_1\Delta_1} \cdots \int_{(k_{|P_c|}-1)\Delta_{|P_c|}}^{k_{|P_c|}\Delta_{|P_c|}} r_p((m_i, x))dx_1, \ldots dx_{|P_c|} \quad (7)$$

for $m_i \in \mathcal{M}_d$ as the flow rate of place $p \in P_c$ when the system is in discrete state $(m_i, k) \in \mathcal{S}_\Delta$. Again the integrals can be evaluated as finite sums since the functions are piecewise constant. For $k = (k_1, \ldots, k_{|P_c|})$ let $k \pm 1_p = (k_1, \ldots, k_{p-1}, k_p \pm 1, k_{p+1}, \ldots, k_{|P_c|})$. Observe that $k + 1_p$ is not defined if $k_p = B(p)$ and $k - 1_p$ is not defined for $k_p = 1$. Now define the flow rates

$$w_i^{k,l} = \begin{cases} \frac{r_{i,k}^p}{\Delta_p} & \text{if } l = k + 1_p \wedge r_{i,k}^p > 0, \\ -\frac{r_{i,k}^p}{\Delta_p} & \text{if } l = k - 1_p \wedge r_{i,k}^p < 0, \\ -\sum_{\ell} \neq k w_i^{k,\ell} & \text{if } l = k, \\ 0 & \text{otherwise,} \end{cases} \quad (8)$$

and the matrix

$$\overline{W}_i = \begin{pmatrix} w_i^{(1,\ldots,1),(1,\ldots,1)} & \cdots & w_i^{(1,\ldots,1),(B(1),\ldots,B(|P_c|))} \\ \vdots & \ddots & \vdots \\ w_i^{(B(1),\ldots,B(|P_c|)),(1,\ldots,1)} & \cdots & w_i^{(B(1),\ldots,P(|P_c|)),(B(1),\ldots,B(|P_c|))} \end{pmatrix}. \quad (9)$$

Then

$$\hat{Q} = \overline{Q} + \overline{W} \text{ where } \overline{W} = \begin{pmatrix} \overline{W}_1 & & \\ & \ddots & \\ & & \overline{W}_{|\mathcal{M}_d|} \end{pmatrix} \quad (10)$$

is the infinitesimal generator matrix of the discretized process such that

$$\frac{du(\tau)}{d\tau} = u(\tau)\hat{Q} \quad (11)$$

is the system of ordinary differential equations describing the evolution of the discretized process. Let $u(\tau)$ be the solution of (11) at time τ starting from $u(0)$, which is the discretized version of $\bar{m}_0$.

To approximate $E(G(\tau))$, we first define the discretized gain vector for continuous places.

$$g_c^k = \int_{(k_1-1)\Delta_1}^{k_1\Delta_1} \cdots \int_{(k_{|P_c|}-1)\Delta_{|P_c|}}^{k_{|P_c|}\Delta_{|P_c|}} g_c(x)dx_1, \ldots dx_{|P_c|}. \quad (12)$$

Then

$$E(G(\tau)) \approx \sum_{m_i \in \mathcal{M}_d} \sum_k u_{(i,k)}(\tau)\left(g_d(m_i) + g_c^k\right), \quad (13)$$

where $\approx$ indicates the inaccuracy by discretization. For later use we define column vectors $\boldsymbol{g}_c, \boldsymbol{g}_d$ of length n_Δ such that $E(G(\tau)) \approx \hat{G}(\tau) \triangleq \boldsymbol{u}(\tau)(\boldsymbol{g}_c + \boldsymbol{g}_d)$.

We denote a discretization $\boldsymbol{\Delta}$ as consistent for $p \in P_c$, if $(b_p^k - b_p^{k-1})/\Delta_p \in \mathbb{N}$ for all intervals. In this case the integrals in (5), (7), (12) can be substituted by sums. A discretization $\boldsymbol{\Delta}$ is consistent if it is consistent for all $p \in P_c$. A refinement of a discretization for place $p \in P_c$ means to substitute Δ_p by Δ_p/i for $i \in \mathbb{N}$. The number of intervals is increased by a factor i. $\boldsymbol{\Delta}/2$ means that every Δ_p is substituted by $\Delta_p/2$. The number of states is of the discrete process is in this case increased by factor $2^{|P_c|}$. If $\boldsymbol{\Delta}$ is consistent, then every refinement is also consistent.

4   Lumping of the Discrete Process

In this section we consider the lumpability of the discretized system developed above. Let $\sim$ be an equivalence relation on S_Δ, $\tilde{S}_\Delta$ the set of equivalence classes and $[\boldsymbol{m}, \boldsymbol{k}]$ the equivalence class to which $(\boldsymbol{m}, \boldsymbol{k}) \in \tilde{S}_\Delta$ belongs. $\sim$ is lumpable relation, iff $\forall \tilde{s}, \tilde{s}' \in \tilde{S}_\Delta, \forall (\boldsymbol{m}_i, \boldsymbol{k}), (\boldsymbol{m}_j, \boldsymbol{l}) \in [\tilde{s}]$:

$$g_c^k = g_c^l, \quad g_d(\boldsymbol{m}_i) = g_d(\boldsymbol{m}_j),$$
$$\sum_{(\boldsymbol{m}_z, \boldsymbol{y}) \in [\tilde{s}']} \hat{Q}((\boldsymbol{m}_i, \boldsymbol{k}), (\boldsymbol{m}_z, \boldsymbol{y})) = \sum_{(\boldsymbol{m}_z, \boldsymbol{y}) \in [\tilde{s}']} \hat{Q}((\boldsymbol{m}_j, \boldsymbol{l}), (\boldsymbol{m}_z, \boldsymbol{y})) \quad (14)$$

The union of lumpable relations is again a lumpable relation. The lumpable relation with the least number of equivalence classes exists and can be defined as the transitive closure of the union of lumpable partitions. In the sequel we denote this relation by $\sim$. It can be computed using partition refinement. Efficient algorithms have been proposed in the past [7,18] and can also be used in our setting.

Let n_Δ be the number of states and $\tilde{n}_\Delta$ the number of equivalence classes of $\sim$. The equivalence relation can be represented by a so-called collector matrix [1] which is a $n_\Delta \times \tilde{n}_\Delta$ matrix $\boldsymbol{V}$ with $\boldsymbol{V}(j, i) = 1$ if $j \in [i]$ and 0 if $j \notin [i]$. Matrix $\boldsymbol{V}$ contains one element equal to 1 in each row and at least one element equal to 1 in each column. A distributor matrix is defined as an $\tilde{n}_\Delta \times n_\Delta$ matrix $\boldsymbol{W} = \overline{(\boldsymbol{V})^T}$ where the overline means that the rows of the transposed matrix $\boldsymbol{V}$ are normalized to 1. Then $\tilde{Q} = \boldsymbol{W}\hat{Q}\boldsymbol{V}$ is the $\tilde{n}_\Delta \times \tilde{n}_\Delta$ matrix of the lumped system. It is easy to show [1,15] that the relation $\hat{Q}\boldsymbol{V} = \boldsymbol{V}\tilde{Q}$ holds in this case. Furthermore, define $\tilde{\boldsymbol{g}}_c^T = \boldsymbol{g}_c^T \boldsymbol{V}$ and $\tilde{\boldsymbol{g}}_d^T = \boldsymbol{g}_d^T \boldsymbol{V}$ which implies, due to the lumpability conditions, $\boldsymbol{g}_c = \boldsymbol{V}\tilde{\boldsymbol{g}}_c$ and $\boldsymbol{g}_d = \boldsymbol{V}\tilde{\boldsymbol{g}}_d$. Then

$$\frac{\partial \tilde{\boldsymbol{u}}(\tau)}{\partial \tau} = \tilde{\boldsymbol{u}}(\tau)\tilde{Q} \quad (15)$$

are the ordinary differential equations for the lumped system. The initial condition is $\tilde{\boldsymbol{u}}(0) = \boldsymbol{u}(0)\boldsymbol{V}$. $E(G(\tau)) \approx \tilde{G}(\tau) = \tilde{\boldsymbol{u}}(\tau)(\tilde{\boldsymbol{g}}_c + \tilde{\boldsymbol{g}}_d)$ is the result of the lumped system.

Theorem 1. *If the lumped system has been generated according to some lumpable equivalence relation $\sim$, then $\tilde{G}(\tau) = \hat{G}(\tau)$.*

Proof. We show here that the forward Euler method with time step δ applied to (11) and (15) yields the same results. Since the Euler method converges for $\delta \to 0$ towards the exact solution, both sets of ordinary differential equations converge towards the same solution. Let $u^k = u(k \cdot \delta)$ and $\tilde{u}^k = \tilde{u}(k \cdot \delta)$, then $\tilde{u}^0 = u^0 V$. We show by induction that the relation holds for all $k = 0, 1, \ldots$.

Assume that $\tilde{u}^k = u^k V$, then the $(k+1)$th vector is computed by the forward Euler scheme as

$$\tilde{u}^{k+1} = \tilde{u}^k + \delta \tilde{u}^k \tilde{Q} = u^k V + \delta u^k V \tilde{Q} = \left(u^k + \delta u^k \hat{Q} \right) V = u^{k+1} V$$

and

$$\tilde{G}(k \cdot \delta) = \tilde{u}^k (\tilde{g}_c + \tilde{g}_d) = u^k V (\tilde{g}_c + \tilde{g}_d) = u^k (g_c + g_d) = \hat{G}(k \cdot \delta),$$

which completes the proof. $\qquad\qquad\qquad\qquad\qquad\qquad\qquad\qquad\qquad\qquad\square$

We now consider refinements of consistent partitions. Let Δ be a partition which is consistent for $p \in P_c$ and let Δ' be a partition that results from Δ by substituting Δ_p by $\Delta_p/2$, then each state $(m, k) \in S_\Delta$ is represented by two states $(m, k^-), (m, k^+) \in S_{\Delta'}$ where $k^- = (k_1, \ldots, k_{p-1}, 2k_p-1, k_{p+1} \ldots, k_{|P_c|})$ and $k^+ = (k_1, \ldots, k_{p-1}, 2k_p, k_{p+1} \ldots, k_{|P_c|})$. The non-diagonal elements of matrix $\hat{Q}_{\Delta'}$ can be derived from the elements of $\hat{Q}_\Delta$ as follows:

$$\begin{aligned}
\hat{Q}_{\Delta'}((m, k^+), (m', k^+)) &= \hat{Q}_{\Delta'}((m, k^-), (m', k^-)) = \hat{Q}_\Delta((m, k), (m', k)) \\
\hat{Q}_{\Delta'}((m, k^+), (m, l^+)) &= \hat{Q}_{\Delta'}((m, k^-), (m, l^-)) = \hat{Q}_\Delta((m, k), (m, l)) \\
\hat{Q}_{\Delta'}((m, k^+), (m, k^-)) &= -\frac{\min(r^p_{m,k}, 0)}{2\Delta_p} \\
\hat{Q}_{\Delta'}((m, k^-), (m, k^+)) &= \frac{\max(r^p_{m,k}, 0)}{2\Delta_p} \\
\hat{Q}_{\Delta'}((m, k^+), (m, (k^+ + 1_p)) &= \frac{\max(r^p_{m,k}, 0)}{2\Delta_p} \\
\hat{Q}_{\Delta'}((m, k^-), (m, (k^- - 1_p)) &= -\frac{\min(r^p_{m,k}, 0)}{2\Delta_p}
\end{aligned}$$

$$(16)$$

for $m \neq m'$ and $l \neq k$ and $l \neq k \pm 1_p$. All remaining non-diagonal matrix entries are 0. The diagonal elements $\hat{Q}_{\Delta'}((m, k^+), (m, k^+)), \hat{Q}_{\Delta'}((m, k^-), (m, k^-))$ are chosen to yield row sum 0 in each row. For the lumped system let $[m, l]$ be the equivalence class to which state (m, l) belongs. Each equivalence class $[m, k]$ is split into two classes $[m, k^-], [m, k^+]$. Matrix $\tilde{Q}_{\Delta'}$ is defined by (16) where $\hat{Q}_{\Delta'}((m, k), (m', k'))$ is substituted by $\hat{Q}_{\Delta'}((m, k), (m', k'))$.

Theorem 2. *Let Δ' be a refinement of Δ. If $\tilde{Q}_\Delta$ results from $\hat{Q}_\Delta$ by a lumpable equivalence relation $\sim$, then $\tilde{Q}_{\Delta'}$ results from $\hat{Q}_{\Delta'}$ by a lumpable equivalence relation $\sim'$ where $(m, k) \sim (m', l) \Rightarrow ((m, k^-) \sim' (m', l^-)) \wedge ((m, k^+) \sim' (m', l^+))$.*

Proof. Consider a pair of states $(\boldsymbol{m}, \boldsymbol{k}^+) \sim' (\boldsymbol{m}', \boldsymbol{l}^+)$. Then it holds that $(\boldsymbol{m}, \boldsymbol{k}) \sim (\boldsymbol{m}', \boldsymbol{l})$. We have to show that the sum of rates into the state from each equivalence class $[\boldsymbol{m}'', \boldsymbol{y}^\pm]$ is identical for both states. All rates out of $(\boldsymbol{m}, \boldsymbol{k}^+)$ $((\boldsymbol{m}', \boldsymbol{l}^+), \text{resp.})$ that do not result from a flow into or out of place p are identical to the corresponding rates out of state $(\boldsymbol{m}, \boldsymbol{k})$ $((\boldsymbol{m}', \boldsymbol{l}), \text{resp.})$. Thus, the rates in the first two cases of (16) are identical, which means that we only have to consider the remaining cases.

Since the discretization is consistent, $r^p_{\boldsymbol{m},\boldsymbol{k}} = r^p_{\boldsymbol{m}',\boldsymbol{l}}$ has to hold which implies that $\frac{r^p_{\boldsymbol{m},\boldsymbol{k}}}{2\Delta'} = \frac{r^p_{\boldsymbol{m}',\boldsymbol{l}}}{2\Delta'}$ and lumpability is transferred from $\sim$ to $\sim'$.

The proof for the other case, $(\boldsymbol{m}, \boldsymbol{k}^-) \sim' (\boldsymbol{m}', \boldsymbol{l}^-)$ uses similar arguments.

$\square$

If $r^p_{\boldsymbol{m},\boldsymbol{k}} = 0$ for all $p \in P_c$, then $\sim'$ can be extended by joining the equivalence classes $[\boldsymbol{m}, \boldsymbol{k}^-]$ and $[\boldsymbol{m}', \boldsymbol{k}^+]$. This can be seen by noticing that the flow between $(\boldsymbol{m}, \boldsymbol{k}^+)$ and $(\boldsymbol{m}, \boldsymbol{k}^-)$ is 0 in this case.

The results presented in this section suggest the following lumping approach:

1. Find the coarsest consistent discretization $\boldsymbol{\Delta}$.
2. Build matrix $\hat{\boldsymbol{Q}}_{\boldsymbol{\Delta}}$.
3. Compute the largest lumpable equivalence relation using partition refinement.
4. Refine the discretization and generate the lumped matrices using (16) such that the discretization error remains small enough. The discretization error for the finite volume method, that is applied, can be estimated using standard methods for the numerical solution of partial differential equations [14].
5. Compute the results by solving the set of ordinary differential equations (11) resulting from the lumped system.

The refinement described here refines Δ_p for all fluid places in the same way. It is, of course, possible to restrict the refinement to some fluid places only.

The whole approach works well, if the transition and flow rates are defined in such a way that a coarse consistent discretization exists. Otherwise the lumpable partition has to be computed for the fine discretization that is used for analysis. Such an approach is only useful for large time horizons τ, for which the solution should be computed.

Example 1. We consider the simple net shown in Fig. 1(a). It describes a source and a sink which have N_d operational modes. In mode $i \in \{0, \ldots, N_d\}$, indicated by i tokens on place p_2 or p_3, fluid is produced respectively consumed. Production of fluid is described by the transitions t_2, t_3, consumption by the transitions t_4, t_5. In mode 0 the consumer or producer are switched off, which means that the transitions are not enabled. The net contains two fluid buffers modeled by the fluid places p_5 and p_6. Both buffers have the same capacity B but buffer p_6 is always filled and emptied first. This implies that the behavior is non-symmetric, the fluid densities differ for both fluid places.

The system becomes lumpable if the input transition t_2 and t_3 as well as the output transitions t_3 and t_4 of the buffers have identical parameters which depend only on the sum of fluid in fluid places and not on the individual fluid

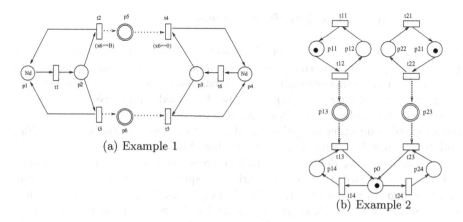

(a) Example 1

(b) Example 2

Fig. 1. Example FSPNs.

levels x_5 and x_6. For the firing rates this means that t_2 is enabled if $m_2 > 0$, $x_5 < B$ and $x_6 = B$. The firing rate is then given by $F(t_2, (\boldsymbol{m}, x_5, x_6)) = \lambda(\boldsymbol{m}, x_5 + x_6)$. Transition t_3 is enabled if $m_2 > 0$ and $x_6 < B$. The firing rate equals $F(t_3, (\boldsymbol{m}, x_5, x_6)) = \lambda(\boldsymbol{m}, x_5 + x_6)$. Similarly, t_4 is enabled with firing rate $F(t_4, (\boldsymbol{m}, x_5, x_6)) = \mu(\boldsymbol{m}, x_5 + x_6)$ if $m_3 > 0, x_5 > 0$ and $x_6 = 0$ and t_5 is enabled with rate $F(t_5, (\boldsymbol{m}, x_5 + x_6)) = \mu(\boldsymbol{m}, x_5 + x_6)$ if $m_3 > 0$ and $x_6 > 0$. $\lambda(.)$ and $\mu(.)$ are positive functions. The flow rates underlie similar restrictions. We have $R((t_2, p_5), (\boldsymbol{m}, x_5, x_6)) = \nu(\boldsymbol{m}, x_5 + x_6)$, $R((t_3, p_6), (\boldsymbol{m}, x_5, x_6)) = \nu(\boldsymbol{m}, x_5 + x_6)$, $R((t_4, p_5), (\boldsymbol{m}, x_5, x_6)) = \xi(\boldsymbol{m}, x_5 + x_6)$ and $R((t_5, p_6), (\boldsymbol{m}, x_5, x_6)) = \xi(\boldsymbol{m}, x_5 + x_6)$. $\nu(.)$ and $\xi(.)$ are in our setting piecewise constant positive functions. Lumping additionally requires that all results are only based on the discrete marking and the sum of the content of the fluid places, $x_5 + x_6$.

If the system is lumpable, in some situations only the sum of fluid of the fluid places and not the individual fluid levels have to be considered in the state space. This results in a significant state space reduction. Table 1 shows some state space sizes for the detailed and lumped system of equations. Parameter N_d describes the number of tokens in the places p_1 and p_4 in the initial marking and N_c is the number of discretization intervals for the fluid places.

Table 1. State space sizes for the first example net.

N_d	N_c	Original	Lumped	N_d	N_c	Original	Lumped
1	6	140	40	2	6	318	93
1	11	480	80	2	11	1083	183
1	21	1760	160	2	21	3963	363
1	51	10400	400	2	51	23403	903
1	101	40800	800	2	101	91803	1803

5   Lumping of the FSPN Matrices

Lumping, as presented in the previous section, is done at the state space of the discrete process. Alternatively several approaches have been proposed to perform lumping compositionally [2,11] or based on symmetries at the net level [4]. In principle similar approaches can be developed for FSPNs after defining compositional or colored nets. However, here we consider an intermediate step by defining and computing lumpable partitions at the levels of the matrices $Q(x)$ and $R_p(x)$ $(p \in P_c)$, defined at the beginning of Sect. 3.

In Theorem 3, we define an equivalence relation $\sim$ that relates states $\bar{m} = (m, x)$ and $\bar{m}' = (m', x')$. We can restrict the equivalence relation to the discrete and continuous parts of the state description. Thus, if $(m, x) \sim (m', x')$, then $m \sim_d m'$ and $x \sim_c x'$. However, in general, the discrete and continuous parts of the relation are not independent. I.e., $(m, x) \sim (m', x') \Rightarrow m \sim_d m' \wedge x \sim_c x'$ but the other direction $\Leftarrow$ usually will not hold. We furthermore assume that $\sim_c$ defines equivalence classes on the set of continuous places. This is a restriction because we consider only continuous states as equivalent which are identical up to the ordering of components in the state vector. Therefore we define permutations $perm$ that permute the indices of continuous places. With a slight misuse of notation we may use $perm$ on the state vector of the continuous part, e.g. $perm(x)$ describes the renumbering of the positions in the vector according to $perm$. If $x \sim_c x'$, then there exists some permutation $perm$ of continuous places, such that $x' = perm(x)$. Thus, $\sim_c$ induces an equivalence relation on P_c and $x \sim_c x'$ implies that x' results from x by reordering equivalent places. We denote by $\mathcal{P}_c$ the set of permutations that permute equivalent continuous places. Since $\sim_p$ is an equivalence relation if $perm \in \mathcal{P}_c$, then $perm^{-1} \in \mathcal{P}_c$ and if $perm, perm' \in \mathcal{P}_c$, then $perm \circ perm' \in \mathcal{P}_c$ where $perm \circ perm'$ is the concatenation of permutations $perm$ and $perm'$.

Theorem 3. *An equivalence relation $\sim$ defines a lumpable partition on the state space of a FSPN, if for all equivalence classes $[m, x], [m', x']$ and for all $(m_i, x_i), (m_j, x_j) \in [m, x]$ the following relations hold:*

$$\exists \text{ a set of permutations } \mathcal{P}_c \text{ on } P_c \text{ such that } x_i = perm(x_j) \text{ for some } perm \in \mathcal{P}_c,$$
$$\sum_{(m_k, x_i) \in [m', x']} q_{ik}(x_i) = \sum_{(m_k, x_j) \in [m', x']} q_{jk}(x_j), \ g_d(m_i) = g_d(m_j), \ g_c(x_i) = g_c(x_j),$$
$$\forall p \in P_c : r_p(m_i, x_i) = r_{perm(p)}(m_j, x_j), B(p) = B(perm(p)) \text{ for } x_j = perm(x_i).$$
$$(17)$$

Proof. We have to prove that (14) holds for arbitrary discretizations Δ where $\Delta_p = \Delta_{p'}$ if $p = perm(p')$ for some $perm \in \mathcal{P}_\sim$.

Now let Δ be some discretization of the above form and $\hat{Q}$ the corresponding rate matrix. $\sim$ induces an equivalence relation $\dot{\sim}$ on the discrete state space such that $(m, k) \dot{\sim} (m, k')$ iff $m \sim m'$ and $k = perm(k')$ for some $perm \in \mathcal{P}_\sim$.

We first consider equivalence of the rewards. For the continuous and discrete reward we have by definition of $\sim$ that $g_d(m_i) = g_d(m_j)$ and $g_c(x_i) = g_c(x_j)$

holds for $m_i \sim m_j$ and $x_i \sim x_j$. For continuous places this implies that also the integrals in (12) are identical for k and k' which implies $g_c^k = g_c^{k'}$.

Now consider the lumpability condition on the sums of rates, namely

$$\sum_{(n',l')\in[(n,l)]} \widehat{Q}((m,k),(n',l')) = \sum_{(n',l')\in[(n,l)]} \widehat{Q}((m',k'),(n',l'))$$

for some equivalence class $[(n,l)]$ of $\sim$ and $(m,k) \sim (m',k')$. The rates can result from a change of the discrete marking by firing a transition (collected in $\overline{Q}$) or from the discretized continuous flow (collected in $\overline{W}$). In the former case the rate for $(m,k) \notin [n,l]$ is given by

$$\sum_{(n',l')\in[(n,k)]} \widehat{Q}((m,k),(n',l')) =$$
$$\sum_{(n',l')\in[(n,l)]} \int_{(k_1-1)\Delta_1}^{k_1\Delta_1} \cdots \int_{(k_{|P_c|}-1)\Delta_{|P_c|}}^{k_{|P_c|}\Delta_{|P_c|}} q_{m,n'}(x)dx_1,\ldots dx_{|P_c|} =$$
$$\int_{(k_1-1)\Delta_1}^{k_1\Delta_1} \cdots \int_{(k_{|P_c|}-1)\Delta_{|P_c|}}^{k_{|P_c|}\Delta_{|P_c|}} \sum_{(n',l')\in[(n,l)]} q_{m,n'}(x)dx_1,\ldots dx_{|P_c|} =$$
$$\sum_{(n',l')\in[(n,l)]} \int_{(k_1-1)\Delta_1}^{k_1\Delta_1} \cdots \int_{(k_{|P_c|}-1)\Delta_{|P_c|}}^{k_{|P_c|}\Delta_{|P_c|}} q_{m',n'}(perm(x))dx_1,\ldots dx_{|P_c|} =$$
$$\sum_{(n',l')\in[(n,l)]} \widehat{Q}((m',k'),(n',l'))$$

Observe that the continuous part is not modified by firing the discrete transition but discretized vectors may differ due to a permutation of equivalent continuous places. Sum and integrals can be interchanged due to Fubini's theorem. If the identity holds for all equivalence classes $[n,l] \neq [m,k]$ then it also holds for $[m,k]$ because $\overline{Q}$ has zero row sums.

If the state changes due to the discretized continuous flow, the discrete remains and we have for the flow according to place $p \in P_c$.

$$\sum_{(n',k)\in[(n,k)]} \widehat{Q}((m,k),(n',k\pm 1_p)) =$$

$$\left| \frac{1}{\Delta_p \prod_{p'\in P_c} \Delta_{p'}} \sum_{(n',l')\in[(n,k\pm 1_p)]} \int_{(k_1-1)\Delta_1}^{k_1\Delta_1} \cdots \int_{(k_{|P_c|}-1)\Delta_{|P_c|}}^{k_{|P_c|}\Delta_{|P_c|}} r_p((m,x))dx_1,\ldots dx_{|P_c|} \right| =$$

$$\left| \frac{1}{\Delta_p \prod_{p'\in P_c} \Delta_{p'}} \int_{(k_1-1)\Delta_1}^{k_1\Delta_1} \cdots \int_{(k_{|P_c|}-1)\Delta_{|P_c|}}^{k_{|P_c|}\Delta_{|P_c|}} \sum_{(n',l')\in[(n,k\pm 1_p)]} r_p((m,x))dx_1,\ldots dx_{|P_c|} \right| =$$

$$\left| \frac{1}{\Delta_p \prod_{p'\in P_c} \Delta_{p'}} \sum_{(n',l')\in[(n,k\pm 1_p)]} \right.$$
$$\left. \int_{(k_1-1)\Delta_1}^{k_1\Delta_1} \cdots \int_{(k_{|P_c|}-1)\Delta_{|P_c|}}^{k_{|P_c|}\Delta_{|P_c|}} r_{perm(p)}((m',perm(x)))dx_1,\ldots dx_{|P_c|} \right| =$$

$$\sum_{(n',k)\in[(n,k)]} \widehat{Q}((m',k'),(n',k'\pm 1_{perm(p)}))$$

$\pm$ equals $+$ if the resulting value is positive, otherwise it equals $-$.

Thus, the lumpability conditions hold for any discretization Δ which completes the proof. $\square$

Partition $\sim$ can be computed by partition refinement. The corresponding algorithm will be briefly outlined in the following steps.

1. Generate the discrete state space $\mathcal{M}_d$ (by assumption this can be done by neglecting continuous places) and set $k = 0$.
2. Define an initial equivalence relation $\sim_d^0$ on $\mathcal{M}_d$ by $\boldsymbol{m} \sim_d^0 \boldsymbol{m}'$ for $\boldsymbol{m}, \boldsymbol{m}' \in \mathcal{M}_d$, iff $g_d(\boldsymbol{m}) = g_d(\boldsymbol{m}')$.
3. Define an initial equivalence relation $\sim_c^0$ by $p \sim_c^0 p'$ for $p, p' \in P_c$ and for all $\boldsymbol{x} \in \mathcal{M}_c$, iff $B(p) = B(p')$ and $g_c(\boldsymbol{x}) = g_c(\boldsymbol{x}_{p \leftrightarrow p'})$ where $\boldsymbol{x}_{p \leftrightarrow p'}$ results from $\boldsymbol{x}$ by exchanging the positions for p and p'. Let $Perm^0$ be the set of all permutations that permute the indices of equivalent places from P_c.
4. Partition refinement of $\sim_d^k$: for all equivalence classes $[\boldsymbol{m}]$ split $[\boldsymbol{m}]$ into new equivalence classes $[\boldsymbol{m}^1], \ldots, [\boldsymbol{m}^L]$ until $\displaystyle\sum_{m_k \in [\boldsymbol{m}']} q_{i,k}(\boldsymbol{x}) = \sum_{m_k \in [\boldsymbol{m}']} q_{j,k}$
   $(perm(\boldsymbol{x}))$ holds for all $\boldsymbol{m}_i, \boldsymbol{m}_j \in [\boldsymbol{m}^l]$ $(l = 1, \ldots, L)$, for all $[\boldsymbol{m}'] \in \sim_d^k$, for all $\boldsymbol{x} \in \mathcal{M}_c$ and some $perm \in Perm^k$, add equivalence classes $[\boldsymbol{m}^1], \ldots, [\boldsymbol{m}^L]$ to $\sim_d^{k+1}$.
5. Partition refinement of $\sim_c^k$: for all equivalence classes $[q]$ of $\sim_c^k$ split $[q]$ into equivalence classes $[q^1], \ldots, [q^K]$ until for all $p, p' \in [q^k]$ $(k = 1, \ldots, K)$, all equivalence classes $[\boldsymbol{m}]$ of $\sim_d^{k+1}$, exist $\boldsymbol{m}_i, \boldsymbol{m}_j \in [\boldsymbol{m}]$ $r_p(\boldsymbol{m}_i, \boldsymbol{x}) = r_{p'}(\boldsymbol{m}_j, perm(\boldsymbol{x}))$ for all $\boldsymbol{x}$ where $perm$ is the permutation that assures in step 4 that $\boldsymbol{m}_i$ and $\boldsymbol{m}_j$ are equivalent add equivalence classes $[q^1], \ldots, [q^K]$ to $\sim_c^{k+1}$.
6. If $\sim_d^k \equiv \sim_d^{k-1}$ and $\sim_c^k \equiv \sim_c^{k-1}$, then stop (a lumpable partition has been found), else set $k = k + 1$ and continue with step 4.

Some remarks should be given for the outlined algorithm. The algorithm eventually terminates because in each iteration at least one new equivalence class is generated for the states in $\mathcal{M}_d$ or the places in P_c and the finest equivalence relation is the identity relation where each equivalence class contains a single discrete state or continuous place, respectively. Since the number of places and the number of discrete markings are finite by assumption, the algorithm will stop. Knowledge of $\sim_d$ and $\sim_c$ is not sufficient to define the lumpable relation, additionally we need the relation between equivalent discrete states and the corresponding permutation of equivalent continuous places (see step 4 of the algorithm). The partition refinement in step 4 requires that the rates are identical for all $\boldsymbol{x} = (x_1, \ldots, x_{|P_c|})$ where $x_p \in [0, B(p)]$. To check this algorithmically an appropriate specification of the rates $q_{ik}(\boldsymbol{x})$ is necessary which is also required for the specification of rates depending on the filling of fluid places.

It should be noted that the lumpability conditions for the discrete and continuous part are not symmetric. For the discrete part we define lumping at the state level of the stochastic process, whereas for the continuous part lumping is defined for symmetric places which is more restrictive. Consequently, a coarser lumpable partition may exist which cannot be found by the outlined algorithm above but can be computed by partition refinement of an adequately discretized process (see e.g. Example 1).

Example 2. The second example is a symmetric FSPN which is shown in Fig. 1(b) in a version with two components. Places with names pki belong to component k. Each component models a switched source that produces fluid for a continuous place $pk3$. A single consumer exists which is idle if a token resides on place $p0$. If transition $tk4$ fires, the consumer changes its state to a state where fluid from place $pk3$ is consumed. If transition $tk3$ fires, the consumer goes back to the idle state. The model can be defined for >2 components in exactly the same way.

To allow state space reduction by lumpability, the components and the consumer have to show a symmetric behavior. We define the corresponding conditions for the case of K components. Let $\mathcal{P}_K$ be the set of all permutations of the numbers 1 through K. We use the notation $perm(\boldsymbol{m})$ and $perm(\boldsymbol{x})$ to indicate the application of permutation $perm$ on the vector which means that the vector components belonging to the corresponding places are exchanged. I.e., if $perm(k) = l$, then pki becomes pli and tki becomes tli. Observe that discrete places are mapped on discrete places and continuous places on continuous places. The following equalities have to hold for all $k, l \in \{1, \ldots, K\}$, all $perm \in \mathcal{P}_K$ where $perm(k) = l$, all transitions tki and all places pki to assure lumpability.

$$F(tki, (\boldsymbol{m}, \boldsymbol{x})) = F(tli, (perm(\boldsymbol{m}, \boldsymbol{x})), \ r_{pki}(\boldsymbol{m}, \boldsymbol{x}) = r_{pli}(perm(\boldsymbol{m}, \boldsymbol{x})),$$
$$g_c^{\boldsymbol{x}} = g_c^{perm(\boldsymbol{x})} \text{ and } g_d(\boldsymbol{m}) = g_d(perm(\boldsymbol{m})).$$

Additionally, the bounds for the continuous places have to be identical and the same discretization has to be applied for all continuous places. Observe that due to an appropriate definition of marking dependent transition rates, the firing of transitions may depend on the filling of continuous places. For example $tk4$ may only have a non-zero rate if $pk3$ includes enough fluid and the rate of $tk3$ may grow if $pk3$ becomes empty. Let n_c be the number of discrete intervals resulting from the discretization of the continuous state of each continuous place. Then $\mathcal{S}_\Delta$ contains $(2K + 1)(2n_c)^K$ states.

A lumpable equivalence relation $\sim$ can be generated by defining two states as equivalent if one can be transformed into the other by a permutation $perm \in \mathcal{P}_K$. Thus, $(\boldsymbol{m}, \boldsymbol{x}) \sim (\boldsymbol{m}', \boldsymbol{x}')$ if $perm(\boldsymbol{m}, \boldsymbol{x}) = (\boldsymbol{m}', \boldsymbol{x}')$ for some $perm \in \mathcal{P}_K$. For the reduced state space, we do not have to distinguish the identity of a component, we only have to consider the number of equivalent components which are in a specific state. This is well known from symmetry exploitation in SPNs [4]. For our example this means that if a token resides at place $p0$, then all components are identical and the number of states is reduced from $(2n_c)^K$ to $\binom{2n_c+K-1}{K}$. If a token resides at some place $pk4$, then the identity of the place is not relevant and the remaining components are identical such that the number of states is reduced from $K(2n_c)^K$ to $2n_c * \binom{2n_c+K-2}{K-1}$. Overall the lumped state space $\tilde{\mathcal{S}}_\Delta$ includes $\binom{2n_c+K-1}{K} + 2n_c * \binom{2n_c+K-2}{K-1}$ states. For larger values of K the reduction becomes significant but the state space still remains large if n_c is large.

6   Conclusions

Lumping proved to be an efficient tool in the analysis of discrete state systems such as CTMCs and SPNs. In this work we extended the concept of lumping to FSPNs which are hybrid (discrete and continuous state) systems. The fundamental approach behind this extension is to map the hybrid system to a discrete state one and apply the available lumping relations for the discrete system.

To pursue this approach we presented a discretization of FSPNs and elaborated on the refinement of the discretization step. We showed that the refinement maintains the lumping relation, which is important for utilizing the fact that asymptotic behavior of the discrete system tends to the hybrid one as the discretization step tends to zero. Additionally we presented an approach where the lumping of the continuous part is based on the symmetry among continuous places which is less general but may be proved by generating only the discrete state space without building the discretized continuous part.

References

1. Buchholz, P.: Exact and ordinary lumpability in finite Markov chains. J. Appl. Probab. **31**, 59–75 (1994)
2. Buchholz, P.: A notion of equivalence for stochastic Petri nets. In: Michelis, G., Diaz, M. (eds.) ICATPN 1995. LNCS, vol. 935, pp. 161–180. Springer, Heidelberg (1995). doi:10.1007/3-540-60029-9_39
3. Chen, D., Hong, Y., Trivedi, K.S.: Second-order stochastic fluid models with fluid-dependent flow rates. Perform. Eval. **49**(1/4), 341–358 (2002)
4. Chiola, G., Dutheillet, C., Franceschinis, G., Haddad, S.: Stochastic well-formed colored nets and symmetric modeling applications. IEEE Trans. Comput. **42**(11), 1343–1360 (1993)
5. Ciardo, G., Nicol, D.M., Trivedi, K.S.: Discrete-event simulation of fluid stochastic Petri nets. IEEE Trans. Softw. Eng. **25**(2), 207–217 (1999)
6. Coxson, P.G.: Lumpability and observability of linear systems. J. Math. Anal. Appl. **99**, 435–446 (1984)
7. Derisavi, S., Hermanns, H., Sanders, W.H.: Optimal state-space lumping in Markov chains. Inf. Process. Lett. **87**(6), 309–315 (2003)
8. Gribaudo, M., Horváth, A.: Fluid stochastic Petri nets augmented with flush-out arcs: a transient analysis technique. IEEE Trans. Softw. Eng. **28**(10), 944–955 (2002)
9. Gribaudo, M., Sereno, M., Horváth, A., Bobbio, A.: Fluid stochastic Petri nets augmented with flush-out arcs: modelling and analysis. Discret. Event Dyn. Syst. **11**(1–2), 97–117 (2001)
10. Gribaudo, M., Telek, M.: Fluid models in performance analysis. In: Bernardo, M., Hillston, J. (eds.) SFM 2007. LNCS, vol. 4486, pp. 271–317. Springer, Heidelberg (2007). doi:10.1007/978-3-540-72522-0_7
11. Hillston, J.: A Compositional Approach to Performance Modelling. Cambridge University Press, Cambridge (1996)
12. Horton, G., Kulkarni, V.G., Nicol, D.M., Trivedi, K.S.: Fluid stochastic Petri nets: theory, applications, and solution techniques. Eur. J. Oper. Res. **105**(1), 184–201 (1998)
13. Iacobelli, G., Tribastone, M., Vandin, A.: Differential bisimulation for a Markovian process algebra. In: Italiano, G.F., Pighizzini, G., Sannella, D.T. (eds.) MFCS 2015. LNCS, vol. 9234, pp. 293–306. Springer, Heidelberg (2015). doi:10.1007/978-3-662-48057-1_23

14. Jasak, H.: Errror analysis and estimation for the finite volume method with applications to fluid flows. Ph.D. thesis, University of London, Department of Mechanical Engineering (1996)
15. Kemeny, J.G., Snell, J.L.: Finite Markov Chains. Springer, Heidelberg (1976)
16. Toth, J., Li, G., Rabitz, H., Tomlin, A.S.: The effect of lumping and expanding on kinetic differential equations. SIAM J. Appl. Math. **57**(6), 1531–1556 (1997)
17. Tschaikowski, M., Tribastone, M.: Exact fluid lumpability in Markovian process algebra. Theor. Comput. Sci. **538**, 140–166 (2014)
18. Valmari, A., Franceschinis, G.: Simple $O(m \log n)$ time Markov chain lumping. In: Esparza, J., Majumdar, R. (eds.) TACAS 2010. LNCS, vol. 6015, pp. 38–52. Springer, Heidelberg (2010). doi:10.1007/978-3-642-12002-2_4

Change Detection of Model Transitions in Proxel Based Simulation of CHnMMs

Dávid Bodnár[⊠], Claudia Krull, and Graham Horton

Institut für Simulation und Graphik,
Otto-von-Guericke-Universität Magdeburg, Magdeburg, Germany
david.bodnar@st.ovgu.de

Abstract. To analyze discrete stochastic models, Virtual Stochastic Sensors were developed at the Otto-von-Guericke-University Magdeburg. This procedure makes it possible to reconstruct the behavior of a broader class of hidden models, like Conversive Hidden non-Markovian Models, in a very efficient way. One assumption of this approach is that the distribution functions, which describe the state changes of the system, are time-homogeneous. However, this assumption is not always true when it comes to real world problems.

To overcome this limitation, the paper presents an algorithm where the concept of Virtual Stochastic Sensors was extended with statistical tests to continuously evaluate the parameters of a Conversive Hidden non-Markovian Model and the current results. If needed, the tests stop the execution of the behavior reconstruction and reevaluate the model based on the current knowledge about the system.

The project showed that detecting the change and adjusting the model is possible during the behavior reconstruction, improving reconstruction accuracy. The method was tested using four types of distribution functions, three of which showed very good results. By using this new algorithm, one is able to construct adaptive models for behavior reconstruction without additional conceptual effort. In this way, loss of modeling accuracy due to abstractions in the modeling process can be balanced. Another possible application appears in the case of long time investigations. The change detection method can be invoked after a given period to reevaluate the system model and make the relevant adjustments if needed.

Keywords: Virtual Stochastic Sensor · Conversive Hidden non-Markovian Model · Drift · Change detection

1 Introduction

Virtual Stochastic Sensor (VSSs) were introduced in [4] and are able to reconstruct the behavior of partially observable processes in discrete stochastic systems. In previous research involving VSSs, the model was always based on previous knowledge about the system, which was considered to be relatively accurate.

© Springer International Publishing AG 2017
N. Thomas and M. Forshaw (Eds.): ASMTA 2017, LNCS 10378, pp. 32–46, 2017.
DOI: 10.1007/978-3-319-61428-1_3

However, in the case of real world applications or long time analysis of bigger systems, not all details can be taken into account. Another source of error is, that processes which require human interaction or intervention cannot be considered to be time-homogeneous. This can lead to serious inaccuracy over time, especially in large systems.

To overcome this limitation, the proxel-based solution algorithm of the Conversive Hidden non-Markovian Models (CHnMMs) was combined with two stochastic tests, the Kolmogorov-Smirnov test (KS test) [6] and the Wald-Wolfowitz runs test [12]. These tests are used to monitor the current state of the reconstructed behavior and evaluate the results. The goal is to detect changes in the system configuration compared to the model, as well as deviations during the execution of the reconstruction. The changes which we will try to detect in this research are parameter shifts in the distribution functions in the hidden part of the system. The type of the distributions is considered to be not changeable. Neither is the actual system state space. This paper is a feasibility analysis based on the master's thesis [2] of one of the authors. The goal of the analysis is to find out whether indication and elimination of such non-obvious changes during the behavior reconstruction are possible or not. It is expected that by using such an algorithm one is able to construct adaptive models. With this, a more realistic connection between the model and the real world can be created and maintained over the lifetime of the model, even with evolving system conditions, without requiring additional effort during the model parametrization.

2   Related and Previous Work

As we live in the world of the Internet of Things (IoT), there is an increasing demand for data to be acquired by sensors. However, there are situations when the required information cannot be measured directly. In these cases, virtual sensors [13] and sensor fusion are used to try to get accurate data on the given system. These sensors use mathematical rules to construct their output. A virtual sensor combined with stochastic models results in a VSS [4] which utilizes stochastic knowledge about a given system. One mathematical model that can be used to model the stochastic relationship between the input data and the results of the VSS is a CHnMM. The proxel-based analysis method can then be used to conduct behavior reconstruction for the VSS in order to acquire information in the form of a statistically relevant estimate of non-measurable system parameters.

In this section, a brief overview will be given of the previous work on VSSs and the needed statistical tests.

2.1   Proxels-Based Analysis

The proxel method [10] was introduced in [7] and is able to construct and analyze the state space of discrete stochastic models represented by Stochastic Petri Net (SPNs) in an efficient and controlled way. A proxel (probability element) is the smallest addressable unit of the reachable state space of the stochastic system at a given moment in time. A proxel is a container which contains all relevant information about a possible state:

$$P_x = (m, \boldsymbol{\tau}, p, t) \tag{1}$$

The stored information are the marking of the current discrete system state (m), the transition age variables ($\boldsymbol{\tau}$), the probability of this discrete state and age combination (p), the current time of the analysis (t) and any additional information which might be relevant for the application like the generating path, utilization of a defined state, etc.

As the analysis runs, the proxel algorithm tracks the possible states and the necessary information from one time step to another. The computation is done in discrete time steps. This means that the time of analysis is discretized with a given granularity and the system is considered to be time homogenous between two time steps. The probability of a possible state change is defined by the Hazard Rate Function (HRF) in Eq. (2), where τ is the age of the current state change. The HRF can be understood as the current rate of the given state change if it did not happen yet.

$$\text{HRF}(\tau) = \frac{\text{PDF}(\tau)}{1 - \text{CDF}(\tau)} \tag{2}$$

During the behavior reconstruction, a proxel holds a possible actual system state with a possible history in the given time step. In the next time step, the possible states are computed from the current state by computing the HRF of the active transitions. Impossible or very unlikely states are pruned away from the proxel tree to dampen state space explosion and maintain acceptable computation time. Using this iterative method, the reachable state space of an SPN is generated in discrete time steps, which can then be analyzed to obtain information such as system performance parameters.

2.2 Conversive Hidden non-Markovian Model

The Hidden Markov Models (HMMs) are well-known statistical models, which assume independent state changes and can be observed through symbol emissions which occur with a given probability. Krull [8] created the so-called Hidden non-Markovian Models (HnMMs), where one can use arbitrary continuous distribution functions and concurrent activities to define connection and dependence between events.

The concept of HnMM in combination with the proxel method can be used for solving the evaluation and decoding tasks for more general models than HMMs. In the case of evaluation one tries to compute the probability that a given observation sequence (trace) was generated by the model. While in the case of decoding one tries to find the most likely state sequence (path) which generated a given observation sequence (trace). The proxel algorithm needs to be modified to not just generate the reachable model state space, but to compare this state space to the observed trace. Only the proxels, which could have produced the trace are retained and analyzed further. The thus modified proxel algorithm results in all possible system paths, in discrete steps, which could have generated the observed trace. Therefore, HnMM combined with a proxel-based solution algorithm can be used for behavior reconstruction of partially observable discrete stochastic systems [8]. This can be used for example for gesture recognition [5].

To increase the efficiency of VSSs, the concept of CHnMMs was introduced and analyzed by Buchholz in [3]. The CHnMM is a subclass of the HnMM, where every state change of the hidden part of the system results in a symbol emission, making state changes easier to track. By using this subclass of the HnMMs, one is able to save a lot of computation time and perform more accurate behavior reconstructions by utilizing a continuous time proxel-based solver algorithm instead of a discrete one. During the feasibility analysis, the project was restricted to CHnMMs to be able to easily analyze and compare the results.

2.3   Statistical Hypothesis Testing

Statistical hypothesis testing [1,6,12] is a group of mathematical methods to observe the significance of a statistically relevant statement, commonly referred as the null hypothesis, by using sample sets and assuming a given certainty that the result of the test is right. As it is a well-known and commonly used tool it will not be discussed in details in this paper. The following two statistical tests were used in the current research.

Kolmogorov-Smirnov Test. The KS test [6] is a well-known non-parametric test, which is used for comparing probability distributions. In our case, the test computes the maximal deviation between an Empirical Distribution Function (EDF) and a theoretical Cumulative Distribution Function(CDF) on a regular grid. If this maximal deviation between them exceeds a given threshold, then the test fails and one is able to say that the given sample was not drawn from the given theoretical distribution function. This test works more accurately if there is a difference [11] in the mean between the given distribution functions. However, in the case of deviation in the variance the test can commit a type II error in rare cases.

Wald-Wolfowitz Runs Test. The Wald-Wolfowitz runs test [12] is another well-known non-parametric test which can be used for validating the randomness of a binary sequence. As it is defined in [12] one can convert a continuous random sequence into a binary sequence by using the median value of the continuous variable and defining the values above and below the median as the same state. In this test, one counts the so-called runs in the binary sequence, which is the number of subsequences where the system stays in the same state. If the number of runs differs significantly from the expected number, then the test fails, indicating that the sequence is not random. This can be used for indicating changes in the variance of the dataset effectively [11].

3   Implementation

In a HnMM the states are connected through transitions, which describe the possible state changes in the system. The firing times of these transitions are characterized with given distribution functions. As already mentioned, the change detection algorithm focuses on the parameters of these distribution functions during the behavior reconstruction.

The basic idea behind the change detection algorithm is a sliding window approach where the already described statistical tests evaluate a given amount of last firing times for every single transition in the model. If a change is detected on a transition, then the execution of the behavior reconstruction is stopped and the new parameters are estimated. After the estimation, the algorithm resets the time to the last valid state of the results and continues the computation.

3.1   Restrictions to the Transitions Behavior

Before diving into the details of the different experiments and the results, the limitations of the project need to be discussed. As already mentioned, the project was limited to the investigation of CHnMMs because of the clearer and easier analysis. The reason is that CHnMMs produce a symbol emission with every state change.

The drift recognition was limited to a change in the parameters of the distribution function because specific machines, processes, and occurrences tend to have a very specific type of distribution. In this way, the assumption was introduced that the expert, constructing the model, is able to guess the type of the distribution accurately. However, during the model construction, one tends to have only a limited amount of information on the system. This knowledge also tends to be stationary, not taking any kind of change into account. Let us assume that one gets information about a production line where extreme precision and high qualification is needed. During the model construction time, a qualified key worker gets sick and is temporarily replaced by a less qualified one. In many cases, the data analyst would not notice this change, but the constructed model and thus the behavior reconstruction might already get inaccurate. Considering this, the drift correction was limited to small, non-obvious changes in the system. Dramatic changes should be easily noticed in most cases during runtime, such as a machine in the production line stopping for a longer period of time.

The goal of the project is therefore to fine tune the system during execution time and prevent a constant need of manual adjustment of the system distribution parameters.

3.2   Change Detection Algorithm

For easier understanding, the change detection algorithm was visualized on the flowchart shown in the Fig. 1.

To be able to use statistical tests for the validation of the model, firstly, one should acquire some data about the system. To do this, a sliding window

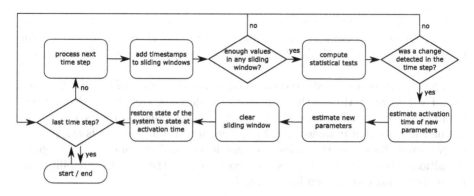

Fig. 1. Flowchart of the change detection algorithm

approach is used. An independent window is defined for all system transitions (state changes) in the model. In these windows, the last n firing times are stored. When one of the windows is filled, the change detection algorithm is activated for the given transition, and the statistical tests determine for the given transition whether a change was detected or not.

The size of this sliding window has a huge impact on the algorithm. If the size is too small, then there might not be enough information present for the statistical tests to perform efficiently. While if the size is too large, then one might get problems with the required memory. Another problem might occur in the case of larger window sizes because the algorithm does not start as long as the sliding window is not filled. This can cause more computations to be performed with the wrong parameters until the algorithm finds the next change.

After the statistical tests are activated, for all active transitions and for every single proxel, the statistical tests are computed. If one of them fails, then a change is indicated for the given transition on the proxel in that timestep. Of course, if every change detected on a single proxel would result in a new estimation of the model, then the computation would never end. That is why a threshold is introduced, which can be seen in the Eq. (3). In the equation, $n_{c,k}$ is the number of proxels where a change was detected in the timestep and n_k is the number of proxels in the timestep. This threshold determines that a change is only recognized as change if the sum of the probabilities of all proxels where a change is detected ($\sum p_{c,i}$) exceeds half of the probabilities of all proxels ($\sum p_l$) in the time step (t_j) normalized with the confidence level ($1 - \alpha$).

$$\frac{\displaystyle\sum_{i=0}^{n_{c,k}} p_{c,i}(t = t_j)}{\displaystyle\sum_{l=0}^{n_k} p_l(t = t_j)} \geq \frac{0.5}{1 - \alpha} \tag{3}$$

If a change is detected, then the algorithm makes a parameter estimation for every single proxel. The estimated parameters (ρ_i) are weighted with the probability of the given proxel (p_i) and normalized. This can be seen in the Eq. (4). Of course, there are also proxels where no change is detected. For these proxels, the original distribution parameters are weighted with the proxel probability.

The algorithm also needs to compute the activation time of the new parameters. The activation time is the time when the detected shift in the distribution parameters likely took place. The computation is similar to the estimation of the new distribution parameters. The unique activation time (δ_i) is assigned to every proxel, based on the oldest element in the sliding window to the given transition. These activation times are weighted with the proxel probability and normalized as it can be seen in the Eq. (5).

$$\rho = \frac{1}{\sum\limits_{l=0}^{n_k} p_l} \sum_{i=0}^{n_k} \rho_i p_i \tag{4}$$

$$\delta = \frac{1}{\sum\limits_{l=0}^{n_k} p_l} \sum_{i=0}^{n_k} \delta_i p_i \tag{5}$$

After the new parameters and the activation time is computed, the algorithm assigns the change to a global array which tracks the activation of these. After that, a reset signal is sent to stop the execution of the behavior reconstruction, go back in time before the activation of the new parameters and recompute that part of the past. To speed up the computation, a ring buffer is implemented which stores the model states periodically. After a reset, only the last valid state needs to be restored and the computation can continue.

3.3 Implementation Challenges

During the implementation, two major challenges were discovered. Distribution functions that only have finite support and a type of aliasing effect, which occurred because of the sliding window approach, causing virtual distributions during the execution. In the following, these will be discussed briefly.

Distribution Functions with Finite Support. Distribution functions with finite support, like a uniform distribution, might cause problems for the change detection algorithm. In these cases the HRF drops to zero if a transition fires outside the strictly defined boundaries of the distribution function. In these cases, the proxel tree would die out and there is no way to continue the analysis.

To overcome this limitation, these distribution functions are altered in that way that a small amount of probability is redistributed from them into newly constructed tails which come from an equivalent normal distribution. The probability of the equivalent normal distribution without the tails is the same as the original finite support distribution without the redistributed probability.

In this way, the HRF will produce small non-zero probabilities outside the support of the original distribution function. Of course, this results in a larger number of proxels with very small probability during the behavior reconstruction, but they are pruned away in the next time step because the probability difference between a firing inside and outside the borders of these newly defined distributions is pronounced.

Virtual Distribution Functions. Virtual distributions occur because the sliding window contains data from two different distribution functions at the same time. Assume that at the beginning of the behavior reconstruction the model is accurate and the sliding windows are filled with samples of the given distribution functions. Then a sudden change occurs. This does not immediately result in a detected change, but when it does, the window still includes samples from the old distribution but has already some from the new one.

As the type of the distribution is assumed to be the same, the parameter estimation would result in a distribution somewhere between the old and the new one with a bit more variance. This is a virtual distribution because it only occurs due to the transient phase. The occurrence of these virtual distributions would not be a problem, if they would not cause an endless loop to happen. The parameters of the virtual distribution cannot fulfill the criteria of the statistical tests. Therefore, the algorithm is not able to leave this state because it always tries to reestimate the same parameters without moving forward.

To overcome this limitation, the algorithm is forced to move forward. After the sliding windows are filled with the required amount of samples, the change detection algorithm does not get activated for a short time. This can be defined as a given amount of time or the number of new symbols which needs to be processed before.

4   Experiments and Results

To make the analysis easy to handle and evaluate, a test scenario was needed. The quality tester example (Figs. 2 and 3) was introduced by Buchholz in [3] and was used during the research in the field of CHnMMs and VSSs. In this example, two independent production lines are merged before a quality tester. The quality tester should indicate when the error rate increases on one of the production lines. It was shown in [3] that this can be done. In our case, we assume that source 0 on the production line is altered in the middle of the analysis (upgrade, maintenance, etc.) and the question is, whether the behavior reconstruction is capable of updating the model accurately and reducing the misclassification error at the same time. The misclassification error denotes the portion of produced items in a trace whose source is incorrectly reconstructed in the path.

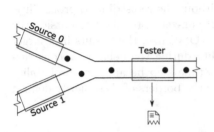
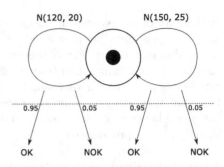

Fig. 2. Visualization of quality tester [3]   **Fig. 3.** Model of the quality tester

4.1 Experiment Setup

In the experiments, two parameters needed to be analyzed because of their significant impact on the results. These were the size of the sliding window and the pruning threshold of the proxel-based solution method.

In the case of the size of the sliding window, as it can be seen in the Fig. 4, a relatively broad range was analyzed between 100 and 250. With small sizes, the algorithm was not robust enough to withstand noise and the transient phase produced a type of numerical oscillation. However, the computation took only 1–10 min. With large sizes, the computation took significantly longer. At the size of 250, it even reached 6 h. In addition, the test set was found to be too short sometimes. A relatively good trade-off between noise and speed can be found around 200–210, where the computation takes about 50–65 min. In our test cases, the amount of 200 timestamps was chosen as window size.

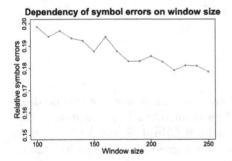

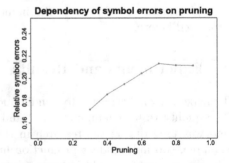

Fig. 4. Relative misclassification error for different window sizes   **Fig. 5.** Relative misclassification error for different pruning thresholds

In our case, we use a relative threshold, meaning that for every time step, all proxel probabilities are compared to the most probable one. If the probability is below a given fraction of this maximum probability, then the proxel is pruned away. This parameter has a huge impact on the computation speed, the needed storage and the amount of significant information gained from the behavior reconstruction. The pruning threshold, as it can be seen in the Fig. 5, was tested between 30% and 90%. The computational times varied between a day and 30 s during these experiments. At 40% the algorithm was executed with an acceptable computation time of about an hour without losing a significant amount of information so during the tests the pruning threshold was held at that level, if possible.

4.2   Results

The experiments were done by using multiple types of distributions with a wide parameter set: Normal, uniform, gamma, and lognormal. Because of the restrictions on the length of this paper, only the results of the test with the normal distributions will be introduced in more detail. For more detailed results on the other distributions, please refer to [2].

The quality criteria for these tests were the following:

- Reconstruction accuracy in terms of misclassification error.
- Accuracy of the distribution parameters for the transition with change.
- Accuracy of the distribution parameters for the transition without change.

In the following heat maps, light green means that there is no error or no change compared to the original value, while red indicates an increase in error/positive parameter deviation and dark blue a decrease in error/negative parameter deviation.

Normal Distribution. The normal distribution is a well-known continuous distribution function often used for representing randomness in examples. The parameters are easy to compute and to estimate, the results are exemplary so that is why it was chosen to represent the results in this paper.

The relative misclassification error needed to be analyzed in two different ways. The results of the first experiment can be seen in the Fig. 6. In this figure, one sees the relative change in the overall misclassified symbols with different distribution parameter changes compared to the case without change detection, when the change happens in the middle of the time protocol. As one can see, there is no significant improvement. In addition to that, in the case of the most dramatic mean change during the analysis, one can notice a decrease in accuracy of about 2–10%. Here the problem lies in the experiment setup. The behavior reconstruction did not spend enough time in the stationary phase at the end of the execution. In this way, there was not enough time to neutralize the effect of the transient phase on the misclassification error.

To have a better overview of this phenomenon, a second experiment was performed. This time, another sequence was generated with the new parameters of the stationary phase and two standard proxel-based solver algorithms were run without the change detection algorithm. In these two runs, the sequence was analyzed with the original and the new model of the HnMM. The difference in the misclassified symbols can be seen in the Fig. 7. The second experiment shows that the algorithm is able to reduce the misclassification error significantly. The

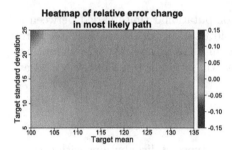

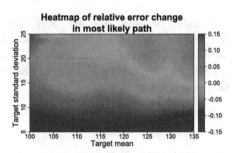

Fig. 6. Relative misclassification error (Color figure online)

Fig. 7. Relative misclassification error if change happens at the start of the behavior reconstruction (Color figure online)

gain in cases with small variance already reaches 10–15%. One can observe an about 2–3% decrease in accuracy in some cases in the upper right corner. In these cases, the two normal distribution functions nearly merge, so the results are nearer to a random guess than to a significant result.

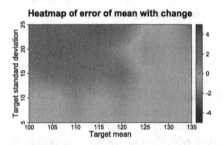

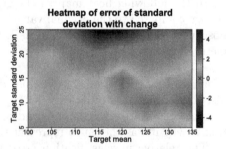

Fig. 8. Error in the mean of the transition with change (Color figure online)

Fig. 9. Error in the standard deviation of the transition with change (Color figure online)

The Figs. 8 and 9 show the error in the distribution parameters at the end of the change detection algorithm. One can notice an error in the upper left corner of the mean diagram. This was caused by the experiment setup. The larger drift in mean with a higher variance had a longer transient phase and some of the computations did not reach the end distribution until the end of the experiment. The same effect can be noticed in the case of the standard deviation in the upper part of the diagram. The red area in the middle of the diagram is an effect of the two distribution functions merging because of the change. Despite these smaller deviations, the results can be considered to be accurate if one takes the relatively short reconstruction time and the drift of the parameters into account. In 75% of the cases, the algorithm performed very accurately in reaching the mean parameter and slight inaccuracies occur only with higher standard deviations. The same accuracy is achieved for the standard deviation in about 50% of the cases.

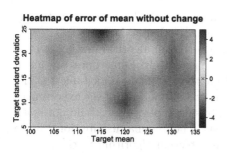

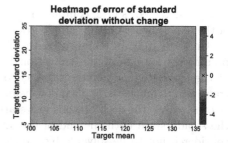

Fig. 10. Error in the mean of the transition without change (Color figure online)

Fig. 11. Error in the standard deviation of the transition without change (Color figure online)

In the Figs. 10 and 11 one can observe the robustness of the algorithm. The noise created by the transient phase indicated some very small change in the transition without change but these changes tend to be compensated in the stationary phase at the end of the behavior reconstruction. The results of the normal distribution show that there is a significant improvement in the misclassification error of the HnMM. The distribution parameters were reached accurately and the transitions are robust enough to withstand the noise of the transient phase accurately.

Lognormal Distribution. The lognormal distribution was chosen as a possible experiment because the parameters are still easy to estimate but the distribution function is not symmetric and some shape change is possible during the change detection. The results were very similar to the ones with the normal distribution.

Uniform Distribution. The uniform distribution was chosen to be an experiment because the reparametrization of distribution functions with finite support needed to be tested. A part of the probability in the Probability Density Function (PDF) was redistributed into the tails as already described. Of course, the HRF computation of the proxel-based solver algorithm was also modified in the same way. The results with the newly constructed distribution exceed the results of the normal distribution when it comes to the accuracy of the parameters. However, it performed slightly worse in the case of the misclassification error.

Gamma Distribution. The gamma distribution is a continuous distribution function with a pronounced shape change depending on the parameters. The gamma distribution also includes the exponential and the chi-squared distribution functions. In this case, the results became very bad. The change detection algorithm almost never reached the required parameters. This was caused by multiple problems. First of all, a more dramatic change in the parameters resulted in a drastic shift of the mean and the variance, which resulted in a longer generated input sequence as the mean got lower. There were much more changes detected, so the computation became very long. This means that some of the experiments on the sample grid ran for more than a day.

Unfortunately, in many cases, the change detection algorithm estimated the parameters for one of the transitions very inaccurately. Sometimes the mean of the estimated parameter set was so high that it made impossible for the transition to fill the sliding window before the end of that part of the experiment. In other cases, the distributions just started to switch places and could not reach the steady state until the end of the analysis. The gamma distribution can be considered not to be solvable by the change detection algorithm in the current stage.

5   Conclusion

5.1   Summary of the Project

The experiments show that the constructed change detection algorithm is able to track a change in the distribution parameters accurately. During the analysis, a gain of about 10–15% in the accuracy of the behavior reconstruction was achieved.

The biggest disadvantage of the algorithm is that it consumes a lot of resources and in some cases takes very long to compute. The memory, required for the computation, jumps from about 50–200 MBs for normal behavior reconstruction to a value of about 2–20 GBs depending on probability difference of the concurrent possible paths. The reason for this huge difference is the implemented ring buffer. Of course, reading and writing chained lists into a ring buffer consumes a lot of time. This is the main reason, that the computation time, required for the algorithm, jumped from a couple of minutes to a value between 10 min and 6 h. Like the required memory, the execution time is very strongly depending on the probability difference of the concurrent possible paths. To overcome this limitation, the change detection algorithm is suggested to be used periodically by very long term behavior reconstructions combined with a general proxel-based solver. By a periodic call of the change detection algorithm, the current model is evaluated and if there is no deviation detected, then the change detection is disabled until the next call to save resources.

Given the results in this paper, a possible change detection algorithm is presented for CHnMMs, despite the inaccurate results in the case of the gamma distribution. By using a more accurate parameter estimator, the algorithm could be a general solution for evaluating the system model during a proxel-based solver algorithm. The new algorithm makes it possible to automatically correct slight inaccuracies in the distribution parameters implemented in the system model unintentionally. Additionally, the algorithm is able to handle model simplifications in a more accurate way by adjusting the system parameters during the behavior reconstruction. In this way, one is able to construct more realistic behavior reconstruction models by using the introduced change detection algorithm.

5.2   Future Research Possibilities

There are multiple possibilities for further development of the algorithm. First of all, a more general and more accurate estimation algorithm might save additional computation time and improve the results of the change detection algorithm. In this feasibility test the project is restricted to CHnMMs. A feasibility test for general HnMM is still needed, which can be built on the basis of the current project. Last but not least, a more efficient ring buffer algorithm might solve the resource shortage. The algorithm should be able to identify key moments during the reconstruction time and save only the states in the ring buffer which are needed in the future with a higher probability. Not just memory consumption can be reduced significantly, but also the computational power of saving and restoring the proxel tree, which would lead to a more acceptable execution time.

5.3   Potential Real World Applications

The algorithm can be used for evaluating the system model in all CHnMMs behavior reconstruction analyses which are long enough to be suitable for the change detection. Of course, this might make the execution time significantly longer, so that the already suggested periodic call might be an option to consider for these cases. The algorithm can also be used for observing the parameter changes of the hidden part of the system model over the execution time. In this way, one should be able to understand the system in a better way and there is a possibility to identify reasons and causes more accurately. The change detection algorithm might be a very good tool to use during behavior reconstruction of production lines and factories like the one in [9]. These processes tend to involve a large amount of human interaction, and this means that changes might occur more often than in other processes. As a result, this change detection algorithm can add a significant gain compared to a standard proxel-based solution algorithm and might help in combination with other information to better understand the processes.

References

1. Bamberg, G., Baur, F., Krapp, M.: Statistik. Oldenbourg Verlag München, München (2011)
2. Bodnár, D.: Change Detection of Model Transitions in Proxel Based Simulation of CHnMMs. Otto-von-Guericke-Universität, Magdeburg (2016)
3. Buchholz, R.: Conversive Hidden non-Markovian Models. Otto-von-Guericke-Universität, Magdeburg (2012)
4. Buchholz, R., Krull, C., Horton, G.: Virtual stochastic sensors: how to gain insight into partially observable discrete stochastic systems. In: The 30th IASTED International Conference on Modelling (2011)
5. Dittmar, T., Krull, C., Horton, G.: A new approach for touch gesture recognition: Conversive Hidden non-Markovian Models. J. Comput. Sci. **10**, 66–76 (2015)
6. Hartung, J., Elpelt, B., Klsener, K.H.: Statistik. Oldenbourg Verlag München, München (2009)

7. Horton, G.: A new paradigm for the numerical simulation of stochastic petri nets with general firing times. In: Proceedings of the European Simulation Symposium, ESS 2002 (2002)
8. Krull, C., Horton, G.: Hidden non-Markovian Models: formalization and solution approaches. In: 6th Vienna International Conference on Mathematical Modelling (2009)
9. Krull, C., Horton, G., Denkena, B., Dengler, B.: Virtual stochastic sensors for reconstructing job shop production workflows. In: 2013 8th EUROSIM Congress on Modelling and Simulation (EUROSIM), pp. 276–281 (2013)
10. Lazarova-Molnar, S.: The Proxel-Based Method: Formalisation, Analysis and Applications. Otto-von-Guericke-Universität, Magdeburg (2005)
11. Magel, R.C., Wibowo, S.H.: Comparing the powers of the Wald-Wolfowitz and Kolmogorov-Smirnov tests. Biom. J. **39**(6), 665–675 (1997)
12. NIST: NIST/SEMATECH e-Handbook of Statistical Methods (2013). http://www.itl.nist.gov/div898/handbook/index.htm
13. Wilson, E.: Virtual sensor technology for process optimization (1997). Presentation at the ISSCAC 1997

Modeling and Analysis of a Relay-Assisted Cooperative Cognitive Network

Ioannis Dimitriou$^{(\boxtimes)}$

Department of Mathematics, University of Patras, 26500 Patras, Greece
`idimit@math.upatras.gr`

Abstract. We investigate a novel queueing system that can be used to model relay-assisted cooperative cognitive networks with coupled relay nodes. Consider a network of two saturated source users that transmit packets towards a common destination node under the cooperation of two relay nodes. The destination node forwards packets outside the network, and each source user forwards its blocked packets to a dedicated relay node. Moreover, when the transmission of a packet outside the network fails, either due to path-loss, fading or due to a hardware/software fault in the transmitter of the destination node, the failed packet is forwarded to a relay node according to a probabilistic policy. In the latter case a recovery period is necessary for the destination node in order to return in an operating mode. Relay nodes have infinite capacity buffers, and are responsible for the retransmission of the blocked/failed packets. Relay nodes have cognitive radio capabilities, and there are fully aware about the state of the other. Taking also into account the wireless interference, a relay node adjusts its retransmission parameters based on the knowledge of the state of the other. We consider a three-dimensional Markov process, investigate its stability, and study its steady-state performance using the theory of boundary value problems. Closed form expressions for the expected delay are also obtained in the symmetrical model.

Keywords: Cooperative network · Cognitive users · Boundary value problem · Stability conditions · Performance

1 Introduction

Nowadays wireless communications technologies have seen a remarkably fast evolution. The new generation of wireless devices has brought notable improvements in terms of communication reliability, data rates, and network connectivity. Furthermore, ad-hoc and sensor networks have emerged with many new applications, where a source has to rely on the assistance from other nodes to forward or relay information to a desired destination [14].

Relay-based cooperative wireless networks have been broadly acknowledged that allows a flexible exchange of data with great benefits on packet delay and energy consumption [6,16]. Such a system operates as follows: There is a finite number of source users that transmit packets to a common destination node, which forwards the received packet outside the network, and a finite number

© Springer International Publishing AG 2017
N. Thomas and M. Forshaw (Eds.): ASMTA 2017, LNCS 10378, pp. 47–62, 2017.
DOI: 10.1007/978-3-319-61428-1_4

of relay nodes that assist source users by retransmitting their blocked/failed packets; e.g., [17,18]. In particular, when a direct source user transmission is blocked, (i.e. the destination node is unavailable), it forwards its blocked packet at a relay node (i.e., a relay overhear the transmission and stores the blocked packet). On the other hand, the transmission of a packet outside the network may also fail for various reasons such as weak radio signals due to distance, multi-path fading, mobility, or faulty device drivers [3]. In such a scenario, the failed packet is also forwarded at a relay node, which retransmits it in a later time instant in order to increase reliability.

Clearly, due to the current trend towards dense wireless networks and the spatial reuse of resources (which in turn increase the impact of interference), it is essential to take into account the interaction among transmissions in the network planning. Moreover, although nowadays there is an increasing demand for variety of wireless applications, the usable radio spectrum is of limited physical extend. Recent studies on the spectrum usage have revealed that substantial portion of the licensed spectrum is underutilized, and thus, there is an imperative need for developing new techniques in order to improve spectrum utilization [13]. The cognitive radio communication is a promising solution to the spectrum underutilization problem [15,18].

Cognitive radio includes a wide range of technologies for making wireless systems "smart". In the full cognitive radio [15] a wireless node is capable to obtain knowledge of its operational environment, and to dynamically reconfigure its operational parameters accordingly. In this direction, we assume that each relay node is capable to exchange information with the other relay node, and thus, it is aware about its state. Therefore, in order to achieve full spectrum utilization of the shared channel, it adjusts its retransmission parameters according to the state of the other relay node (i.e., coupled relay nodes); see also e.g., [4,5,8]. Moreover, in order to avoid further congestion, the source users stop transmitting new packets when they sense the destination node in a recovery mode after a fault at its transmitter. Our system is modeled as a three-dimensional Markov process, and we show that its steady-state performance is expressed in terms of a solution of a boundary value problem; see e.g., [1,2,7–11,20,21].

Our Contribution. Besides its applicability, our work is also theoretically oriented. We provide for the first time, an exact analysis of a continuous time unreliable multiple access cooperative system with queue-aware retransmission control. Our model can be seen as a Markov modulated random walk in the quarter plane (M-RWQP), and its analysis leads to a matrix-form functional equation (i.e., a RWQP modulated by a three-state Markov process). Due to its special structure, we can solve this matrix-form functional equation, and reduce it into a scalar functional equation corresponding to one state of the (modulated) chain. This scalar functional equation is treated using the theory of boundary value problems. To our best knowledge there is no other related work that deals with a detailed analysis of such a system. Ergodicity conditions are also investigated. For the symmetrical system we obtain explicit expressions for the expected delay in each relay node, without solving a boundary value problem. A special case of the proposed model on the traditional cognitive networks is also studied.

The paper is organized as follows. In Sect. 2 we present the model, and form the fundamental functional equation. Ergodicity conditions along with a preparatory mathematical analysis are given in Sect. 3. The generating function of the joint relay queue length distribution for each state of the destination node is given in Sect. 4 by solving two boundary value problems. In Sect. 5, we obtain explicit expressions for the expected delay at each relay node for the symmetrical system. A special case of our model, with potential applications to traditional cognitive networks is investigated Sect. 6. In Sect. 7 we obtain extensive numerical results that show insights into the system performance.

2   The Mathematical Model

We consider a network with two saturated source users, say S_k, $k = 1, 2$, two relay nodes R_1, R_2 and a common destination node D. User S_k generates packets towards node D according to a Poisson process with rate λ_{ki}, $k = 1, 2$[1]. Node D can handle at most one packet, which forwards outside the network. The service time of a packet at node D is exponentially distributed with rate μ.

Moreover, the transmission of a packet outside the network may fail either due to fading/path loss or due to a hardware/software fault in the transmitter of node D. Packet failures due to a hardware/software fault occur according to Poisson process with rate θ. In such a case, node D requires some time to recover. The recovery time is exponentially distributed with rate ν. When the recovery time is completed, the node D returns to its operating state. During recovery time, source users are capable to sense the status of node D (recall that in a cognitive shared access network the nodes exchange information about their status), and in order to avoid further congestion they do not generate packets, i.e., $\lambda_k = 0$, when node D is in the recovery mode.

Cooperation Policy Among Sources and Relays. We assume that relays have infinite capacity buffers and do not generate packets of their own. More precisely, if a transmission of a user's S_k packet to the node D is blocked (i.e., the node D is busy), R_k stores it in its queue and try to forward it later, i.e. S_k cooperates with R_k, $k = 1, 2$ (R_k overhears the transmission, and stores a copy of the blocked packet in its buffer; see [18,19]).

Cooperation Policy Among Node. D and relays. A node's D packet transmission outside the network is successful with probability p (i.e. with probability $q = 1-p$ the transmission fails)[2]. In such a case, the node D cooperates with the relays

---

[1] Note that our analysis is not affected in case we considered more than two source users. However, it cannot be applied when we considered $N > 2$ relay nodes. However, some basic performance metrics and some bounds on the stability conditions can also be obtained; see [9].

[2] The packet success probabilities depend on interference, power etc., and they are commonly determined by the signal to interference plus noise ratio (SINR) threshold model [17,18].

by forwarding a copy of the failed packet according to the following policy: The failed packet is forwarded to R_i with probability r_i, $i = 1, 2$, $r_1 + r_2 = 1$. On the other hand, if a hardware fault occurs at the transmitter of node D, the packet under transmission is either considered lost with probability t_0, or it is forwarded to the R_i with probability t_i, $i = 1, 2$, where $t_0 + t_1 + t_2 = 1$.

Retransmission Policy. In a full cognitive radio network, each relay node is aware of the status of its neighbor, and accordingly, it regulates its retransmission parameters to allow more concurrent communication. Thus, when both relays are non-empty, R_i retransmits a blocked packet to the node D after an exponentially distributed time with rate μ_i, $i = 1, 2$. If R_1 (respectively R_2) empties, then R_2 (respectively R_1) changes its retransmission rate from μ_2 (respectively μ_1) to μ_2^* (respectively μ_1^*).

Let $Q_k(t)$ be the number of packets in relay k, $k = 1, 2$ at time t, and $C(t)$ the state of the node D, where,

$$C(t) = \begin{cases} 0, & \textit{if node D is idle at t,} \\ 1, & \textit{if node D is busy at t,} \\ 2, & \textit{if node D is in recovery state at t,} \end{cases}$$

Clearly, $(Q_1(t), Q_2(t), C(t))$ constitutes a CTMC with state space $E = \{0, 1, \ldots\} \times \{0, 1, \ldots\} \times \{0, 1, 2\}$. Define its stationary probabilities for $(i, j, n) \in E$,

$$p_{i,j}(n) = \lim_{t \to \infty} P(Q_1(t) = i, Q_2(t) = j, C(t) = n) = P(Q_1 = i, Q_2 = j, C = n).$$

Then, for $\lambda = \lambda_1 + \lambda_2$,

$$p_{i,j}(0)[\lambda + \sum\nolimits_{k=1}^{2} \mu_k 1_{\{i,j>0\}} + \mu_1^* 1_{\{i>0,j=0\}} + \mu_2^* 1_{\{i=0,j>0\}}]$$
$$= \mu p p_{i,j}(1) + \nu p_{i,j}(2) + \mu q r_1 p_{i-1,j}(1) 1_{\{i>0\}} + \mu q r_2 p_{i,j-1}(1) 1_{\{j>0\}},$$
$$(\lambda + \mu + \theta) p_{i,j}(1) = \lambda p_{i,j}(0) + \lambda_1 p_{i-1,j}(1) 1_{\{i>0\}} + \lambda_2 p_{i,j-1}(1) 1_{\{j>0\}}$$
$$+ \mu_1 p_{i+1,j}(0) 1_{\{j>0\}} + \mu_1^* p_{i+1,0}(0) 1_{\{j=0\}} + \mu_2 p_{i,j+1}(0) 1_{\{i>0\}} + \mu_2^* p_{0,j+1}(0) 1_{\{i=0\}},$$
$$\nu p_{i,j}(2) = \theta[t_0 p_{i,j}(1) + t_1 p_{i-1,j}(1) 1_{\{i>0\}} + t_2 p_{i,j-1}(1) 1_{\{j>0\}}], \qquad (1)$$

where $1_{\{W\}}$ stands for the indicator function of the event W. Let $H^{(n)}(x, y) = \sum_{i=0}^{\infty} \sum_{j=0}^{\infty} p_{i,j}(n) x^i y^j$, $|x| \le 1$, $|y| \le 1$, $n = 0, 1, 2$. Then, using (1), we obtain,

$$\alpha H^{(0)}(x, y) - (d_2 + \mu_1) H^{(0)}(0, y) - (d_1 + \mu_2) H^{(0)}(x, 0) + (d_1 + d_2)$$
$$\times H^{(0)}(0, 0) = \mu(1 + q(r_1(x - 1) + r_2(y - 1))) H^{(1)}(x, y) + \nu H^{(2)}(x, y), \qquad (2)$$

where $\alpha = \lambda + \mu_1 + \mu_2$. For $d_k = \mu_k - \mu_k^*$, $k = 1, 2$,

$$H^{(0)}(x, y)(\lambda x y + \mu_1 y + \mu_2 x) - x y(\lambda_1(1 - x) + \lambda_2(1 - y)$$
$$+ \mu + \theta) H^{(1)}(x, y) = (d_2 x + \mu_1 y) H^{(0)}(0, y) + (d_1 y + \mu_2 x) H^{(0)}(x, 0)$$
$$- (d_1 y + d_2 x) H^{(0)}(0, 0), \qquad (3)$$
$$H^{(2)}(x, y) = \frac{\theta[1 + t_1(x-1) + t_2(y-1)]}{\nu} H^{(1)}(x, y).$$

Equations (2) and (3) are rewritten in the following matrix-form functional equation,

$$H(x,y)T(x,y) = H(x,0)T_1(x,y) + H(0,y)T_2(x,y) + H(0,0)T_3(x,y), \quad (4)$$

where, $H(x,y) = (H^{(0)}(x,y), H^{(1)}(x,y), H^{(2)}(x,y))$ and

$$T(x,y) = \begin{pmatrix} \alpha & \lambda xy + \mu_1 y + \mu_2 x & 0 \\ -f_0(x,y) & -f_1(x,y) & f_2(x,y) \\ -\nu & 0 & -\nu \end{pmatrix}, \; T_1(x,y) = \begin{pmatrix} d_1 + \mu_2 & d_1 y + \mu_2 x & 0 \\ 0 & 0 & 0 \\ 0 & 0 & 0 \end{pmatrix},$$

$$T_2(x,y) = \begin{pmatrix} d_2 + \mu_1 & d_2 x + \mu_1 y & 0 \\ 0 & 0 & 0 \\ 0 & 0 & 0 \end{pmatrix}, \; T_3(x,y) = \begin{pmatrix} -(d_1 + d_2) & -(d_1 y + d_2 x) & 0 \\ 0 & 0 & 0 \\ 0 & 0 & 0 \end{pmatrix},$$

where,

$$f_0(x,y) = \mu(1 + q(r_1(x-1) + r_2(y-1))),$$
$$f_1(x,y) = xy(\lambda_1(1-x) + \lambda_2(1-y) + \mu + \theta),$$
$$f_2(x,y) = \theta(1 + t_1(x-1) + t_2(y-1)).$$

In general, it is really hard to solve the matrix functional Eq. (4). In the following, we are going to exploit its special structure, and convert (4) into a scalar functional equation. Indeed, using (2) and (3), we obtain,

$$R(x,y)H^{(0)}(x,y) = A(x,y)H^{(0)}(x,0) + B(x,y)H^{(0)}(0,y) + C(x,y)H^{(0)}(0,0), \quad (5)$$

where,

$$\begin{aligned} R(x,y) &= xy(\tilde{\lambda}_2(y-1) + \tilde{\lambda}_1(x-1)) + \mu_1 y[(\mu(p+qr_2) \\ &\quad + \theta(t_0 + t_2))(1-x) + (\theta t_2 + \mu qr_2)(y-1)] \\ &\quad + \mu_2 x[(\mu(p+qr_1) + \theta(t_0 + t_1))(1-y) + (\theta t_1 + \mu qr_1)(x-1)], \end{aligned} \quad (6)$$

$$\begin{aligned} A(x,y) &= \mu_2 x[(1-y)(\mu(p+qr_1) + \theta(t_0 + t_1) - \lambda_2 y) + (x-1)(\theta t_1 + \mu qr_1 + \lambda_1 y)] \\ &\quad + d_1 y[(1-x)(\mu(p+qr_2) + \theta(t_0 + t_2) - \lambda_1 x) + (y-1)(\theta t_2 + \mu qr_2 + \lambda_2 x)], \end{aligned}$$

$$\begin{aligned} B(x,y) &= d_2 x[(1-y)(\mu(p+qr_1) + \theta(t_0 + t_1) - \lambda_2 y) + (x-1)(\theta t_1 + \mu qr_1 + \lambda_1 y)] \\ &\quad + \mu_1 y[(1-x)(\mu(p+qr_2) + \theta(t_0 + t_2) - \lambda_1 x) + (y-1)(\theta t_2 + \mu qr_2 + \lambda_2 x)], \end{aligned}$$

$$\begin{aligned} C(x,y) &= -d_1 y[(1-x)(\mu(p+qr_2) + \theta(t_0 + t_2) - \lambda_1 x) + (y-1)(\theta t_2 + \mu qr_2 + \lambda_2 x)] \\ &\quad - d_2 x[(1-y)(\mu(p+qr_1) + \theta(t_0 + t_1) - \lambda_2 y) + (x-1)(\theta t_1 + \mu qr_1 + \lambda_1 y)]. \end{aligned}$$

where $\tilde{\lambda}_k = \lambda_k(\lambda + \mu_1 + \mu_2 + \theta t_k + \mu qr_k)$, $k = 1, 2$. Thus, our aim in the following is to solve (5) and obtain $H^{(0)}(x,y)$. Then, by substituting back in (3) we can obtain $H^{(1)}(x,y)$, $H^{(2)}(x,y)$, and the vector $H(x,y)$ has been fully determined.

Remark: To our best knowledge, there is no other work in the related literature where such a type of kernel $R(x,y)$ arises.

3   Basic Analysis

In the following, we proceed with the derivation of the ergodicity conditions, and the investigation of basic properties of the kernel equation $R(x,y) = 0$.

3.1 Ergodicity Conditions

Assume in the following that $\lambda + \mu + \mu_1 + \mu_2 + \theta = 1$[3]. Then, (5) is rewritten as

$$R(x,y)\pi^{(0)}(x,y) = \frac{R(x,y)-A(x,y)}{y}\pi_1^{(0)}(x) + \frac{R(x,y)-B(x,y)}{x}\pi_2^{(0)}(y) \\ + \frac{R(x,y)-C(x,y)-A(x,y)-B(x,y)}{xy}p_{0,0}(0), \quad (7)$$

where for $|x| \leq 1$, $|y| \leq 1$, $\pi^{(0)}(x,y) := \sum_{i=1}^{\infty}\sum_{j=1}^{\infty}p_{i,j}(0)x^{i-1}y^{j-1}$, $pi_1^{(0)}(x) := \sum_{i=1}^{\infty}p_{i,0}(0)x^{i-1}$, $\pi_2^{(0)}(y) := \sum_{j=1}^{\infty}p_{0,j}(0)y^{j-1}$. Equation (7) is the fundamental form corresponding to a RWQP whose transition diagram is depicted in Fig. 1. Its one-step transition probabilities are given by

$$\hat{p}_{1,0} = \tilde{\lambda}_1, \hat{p}_{0,1} = \tilde{\lambda}_2, \hat{p}_{-1,0} = \mu_1(\mu p + \theta t_0), \hat{p}_{0,-1} = \mu_2(\mu p + \theta t_0),$$
$$\hat{p}_{1,-1} = \mu_2(\mu q r_1 + \theta t_1), \hat{p}_{-1,1} = \mu_1(\mu q r_2 + \theta t_2),$$
$$\hat{p}_{0,0} = 1 - (\tilde{\lambda} + \mu_1(\mu(p + q r_2) + \theta(t_0 + t_2)) + \mu_2(\mu(p + q r_1) + \theta(t_0 + t_1))),$$
$$\hat{p}_{1,0}^{(1)} = \lambda_1(\lambda + \mu_1^* + \theta t_1 + \mu q r_1), \hat{p}_{0,1}^{(1)} = \lambda_2(\lambda + \mu_1^* + \theta t_2 + \mu q r_2),$$
$$\hat{p}_{-1,0}^{(1)} = \mu_1^*(\mu p + \theta t_0), \hat{p}_{-1,1}^{(1)} = \mu_1^*(\mu q r_2 + \theta t_2), \hat{p}_{0,0}^{(1)} = 1 - (\lambda_1(\lambda + \mu_1^* + \theta t_1 + \mu q r_1) \\ + \lambda_2(\lambda + \mu_1^* + \theta t_2 + \mu q r_2) + \mu_1^*(\mu(p + q r_2) + \theta(t_2 + t_0))),$$

$$\hat{p}_{1,0}^{(2)} = \lambda_1(\lambda + \mu_2^* + \theta t_1 + \mu q r_1), \hat{p}_{0,1}^{(1)} = \lambda_2(\lambda + \mu_2^* + \theta t_2 + \mu q r_2),$$
$$\hat{p}_{0,-1}^{(2)} = \mu_2^*(\mu p + \theta t_0), \hat{p}_{1,-1}^{(1)} = \mu_2^*(\mu q r_1 + \theta t_1), \hat{p}_{0,0}^{(2)} = 1 - (\lambda_1(\lambda + \mu_2^* + \theta t_1 + \mu q r_1) \\ + \lambda_2(\lambda + \mu_2^* + \theta t_2 + \mu q r_2) + \mu_2^*(\mu(p + q r_1) + \theta(t_1 + t_0))),$$
$$\hat{p}_{1,0}^{(0)} = \lambda_1(\lambda + \theta t_1 + \mu q r_1), \hat{p}_{0,1}^{(0)} = \lambda_2(\lambda + \theta t_2 + \mu q r_2),$$
$$\hat{p}_{0,0}^{(0)} = 1 - (\lambda_1(\lambda + \theta t_1 + \mu q r_1) + \lambda_2(\lambda + \theta t_2 + \mu q r_2)).$$

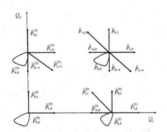

Fig. 1. Transition diagram of the RWQP corresponding to the idle states of node D.

Set, $M = (M_x, M_y) = (\sum_{i,j} i\hat{p}_{i,j}, \sum_{i,j} j\hat{p}_{i,j})$, $(M_x^{(1)}, M_y^{(1)}) = (\sum_{i,j} i\hat{p}_{i,j}^{(1)}, \sum_{i,j} j\hat{p}_{i,j}^{(1)})$, $(M_x^{(2)}, M_y^{(2)}) = (\sum_{i,j} i\hat{p}_{i,j}^{(2)}, \sum_{i,j} j\hat{p}_{i,j}^{(2)})$. Then,

$$M_x = \tilde{\lambda}_1 + \mu_2(\mu q r_1 + \theta t_1) - \mu_1(\mu(p + q r_2) + \theta(t_0 + t_2)),$$
$$M_y = \tilde{\lambda}_2 + \mu_1(\mu q r_2 + \theta t_2) - \mu_2(\mu(p + q r_1) + \theta(t_0 + t_1)),$$
$$M_x^{(1)} = \lambda_1(\lambda + \mu_1^* + \mu q r_1 + \theta t_1) - \mu_1^*(\mu(p + q r_2) + \theta(t_0 + t_2)),$$
$$M_y^{(1)} = \lambda_2(\lambda + \mu_1^* + \mu q r_2 + \theta t_2) + \mu_1^*(\mu q r_2 + \theta t_2),$$

---

[3] A technical assumption to avoid normalizing the one step transition probabilities.

$$M_x^{(2)} = \lambda_1(\lambda + \mu_2^* + \mu q r_1 + \theta t_1) + \mu_2^*(\mu q r_1 + \theta t_1),$$
$$M_y^{(2)} = \lambda_2(\lambda + \mu_2^* + \mu q r_2 + \theta t_2) - \mu_2^*(\mu(p + q r_1) + \theta(t_0 + t_1)).$$

Theorem 1 *(see [11]). When $M \neq 0$, our system is stable if, and only if, one of the following conditions holds,*

1.
$$M_x < 0, M_y < 0, \quad \begin{cases} \Gamma_1 = M_x M_y^{(1)} - M_y M_x^{(1)} < 0, \\ \Gamma_2 = M_y M_x^{(2)} - M_x M_y^{(2)} < 0, \end{cases}$$

2. $M_x < 0$, $M_y \geq 0$, $\Gamma_2 < 0$;
3. $M_x \geq 0$, $M_y < 0$, $\Gamma_1 < 0$.

3.2   The Algebraic Curve $R(x, y) = 0$

Equation (5) includes three unknown functions: $H^{(0)}(x, y)$, and two functions of one complex variable $H^{(0)}(x, 0)$, $H^{(0)}(0, y)$, which we would like to determine. By investigating the kernel equation $R(x, y) = 0$, allows us to come up with a functional equation involving only $H^{(0)}(x, 0)$, $H^{(0)}(0, y)$. Note that,

$$R(x, y) = a(x)y^2 + b(x)y + c(x) = \widehat{a}(y)x^2 + \widehat{b}(y)x + \widehat{c}(y),$$
$$a(x) = \tilde{\lambda}_2 x + \mu_1(\theta t_2 + \mu q r_2), \quad c(x) = \mu_2 x(\mu p + \theta t_0 + (\mu q r_1 + \theta t_1)x),$$
$$b(x) = \tilde{\lambda}_1 x^2 - x[\tilde{\lambda} + \mu_1(\mu(p + q r_2) + \theta(t_0 + t_2))$$
$$+ \mu_2(\mu(p + q r_1) + \theta(t_0 + t_1))] + \mu_1(\mu p + \theta t_0),$$
$$\widehat{a}(y) = \tilde{\lambda}_1 y + \mu_2(\theta t_1 + \mu q r_1), \quad \widehat{c}(y) = \mu_1 y(\mu p + \theta t_0 + (\mu q r_2 + \theta t_2)y),$$
$$\widehat{b}(y) = \tilde{\lambda}_2 y^2 - y[\tilde{\lambda} + \mu_1(\mu(p + q r_2) + \theta(t_0 + t_2))$$
$$+ \mu_2(\mu(p + q r_1) + \theta(t_0 + t_1))] + \mu_2(\mu p + \theta t_0),$$

where $\tilde{\lambda} = \tilde{\lambda}_1 + \tilde{\lambda}_2$. Denote by $X_\pm(y)$, $Y_\pm(x)$ the roots of $R(X(y), y) = 0$, $x \in C_x$ and $R(x, Y(x)) = 0$, $y \in C_y$ respectively, where C_x, C_y the complex planes of x, y, respectively.

To ensure the continuity of $Y(x)$, $X(y)$ let the cut planes, $\tilde{\tilde{C}}_x = C_x - ([x_1, x_2] \cup [x_3, x_4]$, $\tilde{\tilde{C}}_y = C_y - ([y_1, y_2] \cup [y_3, y_4])$. In $\tilde{\tilde{C}}_x$ (resp. $\tilde{\tilde{C}}_y$), denote by $Y_0(x)$ (resp. $X_0(y)$) the zero of $R(x, Y(x)) = 0$ (resp. $R(X(y), y) = 0$) with the smallest modulus, such that $|Y_0(x)| < 1$ (resp. $|X_0(y)| < 1$). In order to proceed, we have to allocate the branch points of $Y(x)$, $X(y)$ (recall that in such points $Y(x)$, $X(y)$ are not regular functions). Using Lemma 2.3.8, pp. 26–27 in [11], and the probabilities $\widehat{p}_{ij}$ of the jumps in the interior of the quarter plane (see Subsect. 3.1) we can easily show that $Y(x)$ (resp. $X(y)$) has four real branch points, say x_i, $i = 1, 2, 3, 4$ such that $0 < x_1 \leq x_2 < 1 < x_3 < x_4 < \infty$ (resp. y_i, $i = 1, 2, 3, 4$ such that $0 < y_1 \leq y_2 < 1 < y_3 < y_4 < \infty$).

Lemma 1

1. $X(y)$, $y \in [y_1, y_2]$ lies on a simple closed contour $\mathcal{M}$ defined by

$$|x|^2 = m(Re(x)), \quad m(\delta) = \frac{\mu_1 \zeta(\delta)[\mu p + \theta t_0 + (\mu q r_2 + \theta t_2)\zeta(\delta)]}{\tilde{\lambda}_1 \zeta(\delta) + \mu_2(\theta t_1 + \mu q r_1)}, \quad |x|^2 \leq \frac{\widehat{c}(y_2)}{\widehat{a}(y_2)},$$

where $\zeta(\delta) = \frac{\widehat{b} - 2\widetilde{\lambda}_1\delta - \sqrt{(\widehat{b} - 2\widetilde{\lambda}_1\delta)^2 - 4\widetilde{\lambda}_2\mu_2(2\delta(\theta t_1 + \mu q r_1) - \mu p - \theta t_0)}}{2\lambda_2}$, $\widehat{b} = \widetilde{\lambda} + \mu_1(\mu(p +$ $qr_2) + \theta(t_0 + t_2)) + \mu_2(\mu(p + qr_1) + \theta(t_0 + t_1))$. Set $\beta_1 = \sqrt{\widehat{c}(y_2)/\widehat{a}(y_2)}$, and $\beta_2 = -\sqrt{\widehat{c}(y_1)/\widehat{a}(y_1)}$, its extreme right and left point respectively.

2. $Y(x)$, $x \in [x_1, x_2]$, lies on a simple closed contour $\mathcal{L}$ defined by

$$|y|^2 = k(Re(y)), \quad k(\delta) = \frac{\mu_2\eta(\delta)[\mu p + \theta t_0 + (\mu q r_1 + \theta t_1)\eta(\delta)]}{\lambda_2\eta(\delta) + \mu_1(\theta t_2 + \mu q r_2)}, \quad |y|^2 \le \frac{c(x_2)}{a(x_2)},$$

where $\eta(\delta) = \frac{\widehat{b} - 2\widetilde{\lambda}_2\delta - \sqrt{(\widehat{b} - 2\widetilde{\lambda}_2\delta)^2 - 4\widetilde{\lambda}_1\mu_1(2\delta(\theta t_2 + \mu q r_2) - \mu p - \theta t_0)}}{2\lambda_1}$. Let $\psi_1 = \sqrt{c(x_2)/a(x_2)}$, $\psi_2 = -\sqrt{c(x_1)/a(x_1)}$, its extreme right and left point, respectively.

Proof: We focus on 2. (1. is proved similarly). Clearly, $Re(Y(x)) = \frac{-b(x)}{2a(x)}$. Solving the previous expression for x with $\delta = Re(Y(x))$, and taking the solution such that $x \in [0,1]$, we obtain $\eta(\delta)$. Moreover, $|Y(x)|^2 = |y|^2 = \frac{c(x)}{a(x)} = h(x)$. Note that $h(x)$ is an increasing function in x. Therefore, $|Y(x)| \le |Y(x_2)| = \psi_1$, which is the extreme right point of $\mathcal{L}$. Similarly, we can show that, $Y_0(x_1) = \psi_2$ is the extreme left point of $\mathcal{L}$. $\qquad\square$

4   Solution of the Fundamental Functional Equation

We firstly proceed with the derivation of some useful relations. Let $p_{i,.}(n) = \sum_{j=0}^{\infty} p_{i,j}(n), i = 0, 1, \ldots, p_{.,j}(n) = \sum_{i=0}^{\infty} p_{i,j}(n), j = 0, 1, \ldots, n = 0, 1, 2$.

Lemma 2

$$H^{(1)}(1,1) = \frac{\lambda}{\mu p + \theta(1 + \frac{\lambda}{\nu})}, \quad H^{(2)}(1,1) = \frac{\theta}{\nu}H^{(1)}(1,1),$$
$$H^{(0)}(1,1) = 1 - H^{(1)}(1,1)(1 + \frac{\theta}{\nu}). \tag{8}$$

1. If $\mu_1\mu_2 \ne d_1 d_2$, then, for $\widehat{\rho}_k = \frac{\lambda(\lambda_k + \mu_k + \mu r_k)}{\mu_k(\mu p + \theta(1 + \lambda/\nu))}, k = 1, 2$,

$$H^{(0)}(0,1) = \frac{\mu_1\mu_2[1 - \widehat{\rho}_1 - \frac{d_1}{\mu_1}(1 - \widehat{\rho}_2)] + d_1\mu_2^* H^{(0)}(0,0)}{\mu_1\mu_2 - d_1 d_2},$$
$$H^{(0)}(1,0) = \frac{\mu_1\mu_2[1 - \widehat{\rho}_2 - \frac{d_2}{\mu_2}(1 - \widehat{\rho}_1)] + d_2\mu_1^* H^{(0)}(0,0)}{\mu_1\mu_2 - d_1 d_2}. \tag{9}$$

2. If $\mu_1\mu_2 = d_1 d_2$, then $\mu_i = \phi_i\mu_i^*$, $i = 1, 2$, $\phi_1 + \phi_2 = 1$, and $H^{(0)}(0,0) = 1 - (\phi_1\widehat{\rho}_1 + \phi_2\widehat{\rho}_2) = 1 - \rho^*$.

Proof: By considering the cut between $\{Q_1 = i, C = 1\}$ and $\{Q_1 = i + 1, C = 0\}$,

$$(\lambda_1 + \mu q r_1)p_{i,.}(1) = \mu_1^* p_{i+1,0}(0) + \mu_1 \sum_{j=1}^{\infty} p_{i+1,j}(0), i \ge 0.$$

Summing for all $i = 0, 1, \ldots$, we derive after some algebra,

$$\lambda_{0,1}H^{(1)}(1,1) = \mu_1[H^{(0)}(1,1) - H^{(0)}(0,1)] - d_1[H^{(0)}(1,0) - H^{(0)}(0,0)], \tag{10}$$

where $\lambda_{0,j} = \lambda_j + \mu q r_j$, $j = 1, 2$. Similarly,

$$\lambda_{0,2} H^{(1)}(1,1) = \mu_2 [H^{(0)}(1,1) - H^{(0)}(1,0)] - d_2 [H^{(0)}(0,1) - H^{(0)}(0,0)]. \quad (11)$$

Note that (10) and (11) are both "conservation of flow" relations. Summing (10) and (11), and subtracting the sum from (2) we can obtain after some algebra Eq. (8). Substituting (8) in (10), (11), we obtain

$$\begin{aligned}
H^{(0)}(0,1) &= 1 - \hat{\rho}_1 - \tfrac{d_1}{\mu_1}[H^{(0)}(1,0) - H^{(0)}(0,0)], \\
H^{(0)}(1,0) &= 1 - \hat{\rho}_2 - \tfrac{d_2}{\mu_2}[H^{(0)}(0,1) - H^{(0)}(0,0)].
\end{aligned} \quad (12)$$

From (12) and for $\mu_1\mu_2 \neq d_1 d_2$ we derive (9). On the other hand, for $\mu_1\mu_2 = d_1 d_2$ (i.e., $\mu_k = \phi_k \mu_k^*$, $k = 1, 2$, $\phi_1 + \phi_2 = 1$), (12) yields $H^{(0)}(0,0) = 1 - \rho^*$. $\quad \square$

Based on Lemma 2, and in particular on the value of $\mu_1\mu_2 - d_1 d_2$ we discriminate the analysis in two cases.

4.1   The General Case

Let $\mu_1\mu_2 \neq d_1 d_2$, and denote

$$\begin{aligned}
H(x) &:= H^{(0))}(x,0) + \frac{d_2\mu_1^*}{\mu_1\mu_2 - d_1 d_2} H^{(0)}(0,0), \\
G(y) &:= H^{(0))}(0,y) + \frac{d_1\mu_2^*}{\mu_1\mu_2 - d_1 d_2} H^{(0)}(0,0).
\end{aligned}$$

For a zeropair $(X_0(y), y)$ of the kernel where $|y| \leq 1$, $|X_0(y)| \leq 1$, (5) becomes

$$A(X_0(y), y) H(X_0(y)) = -B(X_0(y), y) G(y). \quad (13)$$

Using results from Subsect. 3.2 we conclude in,

$$A(x, Y_0(x)) H(x) = -B(x, Y_0(x)) G(Y_0(x)), \quad x \in \mathcal{M}. \quad (14)$$

From (14), we can easily see that the possible poles of $H(x)$, $x \in S_x := G_\mathcal{M} \cap \bar{D}_x^c$ are necessarily the zeros of $A(x, Y_0(x))$ in S_x, where $G_\mathcal{U}$ be the interior domain bounded by $\mathcal{U}$, and $\bar{D}_x^c = \{x : |x| > 1\}$ (using simple algebraic arguments we can show that under stability conditions $H(x)$ has no poles in S_x. Due to space constrains the details are omitted). Thus, $H(x)$ is regular in $G_\mathcal{M}$, continuous in $\mathcal{M} \cup G_\mathcal{M}$, and $U(x) = A(x, Y_0(x))/B(x, Y_0(x))$ is a non-vanishing function on $\mathcal{M}$. Therefore, (14) yields (following [1], $B(x, Y_0(x)) \neq 0$, $x \in \mathcal{M}$),

$$Re[i\tfrac{A(x,Y_0(x))}{B(x,Y_0(x))} H(x)] = 0, \ x \in \mathcal{M}, \quad (15)$$

The problem formulated in (15), must be conformally transformed to the unit circle $\mathcal{C}$. Let $z = f(x) : G_\mathcal{M} \to G_\mathcal{C}$ be the conformal mapping and $x = f_0(z) : G_\mathcal{C} \to G_\mathcal{M}$ its inverse. Upon applying the conformal mapping, our aim is to find a function $F(z) := H(f_0(z))$, regular in $G_\mathcal{C}$, continuous in $G_\mathcal{C} \cup \mathcal{C}$ such that, $Re[iU(f_0(z))F(z)] = 0$, $z \in \mathcal{C}$. The following lemma is based on [11], and provides information about the solvability of (15), based on the value of the index χ of $U(x)$, $x \in \mathcal{M}$ (see [12]).

Lemma 3

1. If $M_y < 0$, then $\chi = 0$ is equivalent to

$$\frac{dA(x,Y_0(x))}{dx}\Big|_{x=1} < 0 \Leftrightarrow \Gamma_1 < 0, \quad \frac{dB(X_0(y),y)}{dy}\Big|_{y=1} < 0 \Leftrightarrow \Gamma_2 < 0.$$

2. If $M_y \geq 0$, $\chi = 0 \Leftrightarrow \frac{dB(X_0(y),y)}{dy}\Big|_{y=1} < 0 \Leftrightarrow \Gamma_2 < 0.$

Thus, under ergodicity conditions, $\chi = 0$, and (15) has a unique solution:

$$H^{(0)}(x,0) = W \exp\left[\frac{1}{2\pi i} \int_{|t|=1} \frac{\log\{J(t)\}}{t - f(x)} dt\right] + \frac{d_2 \mu_1^*}{d_1 d_2 - \mu_1 \mu_2} H^{(0)}(0,0), \tag{16}$$

where W is an unknown constant, and $J(t) = \frac{\overline{U(t)}}{U(t)}$, $U(t) = U(f_0(t))$. Since $1 \in G_{\mathcal{M}}$, W is found as a function of $H^{(0)}(0,0)$ using (16) for $x = 0$. Then, setting in (16) $x = 1$, and combining with (9), we obtain $H^{(0)}(0,0)$. To conclude,

$$W = \frac{\mu_1 \mu_2^*}{\mu_1 \mu_2 - d_1 d_2} \exp\left[\frac{1}{2\pi i} \int_{|t|=1} \frac{\log\{J(t)\}}{t} dt\right] H^{(0)}(0,0),$$
$$H^{(0)}(0,0) = \frac{\mu_2[1 - \hat{\rho}_2 - \frac{d_2}{\mu_2}(1 - \hat{\rho}_1)] \exp\left[\frac{-1}{2i\pi} \int_{|t|=1} \frac{\log\{J(t)\}f(1)}{t(t - f(1))} dt\right]}{\mu_2^*}. \tag{17}$$

Similarly, $H^{(0)}(0,y)$ is obtained by solving another Riemann-Hilbert boundary value problem on $\mathcal{L}$. Then, substituting back in (5), (2), (3) we obtain $H^{(0)}(x,y)$, $H^{(1)}(x,y)$, and $H^{(2)}(x,y)$, respectively.

4.2 The Simple Case

Let, $\mu_1 \mu_2 = d_1 d_2$, i.e., $\mu_k = \phi_k \mu_k^*$, $k = 1, 2$, $\phi_1 + \phi_2 = 1$. Then, $B(x,y) = -\frac{\phi_1}{\phi_2} A(x,y)$. Following the lines in Subsect. 4.1, and taking a zeropair $(x, Y_0(x))$ of the kernel where $|x| \leq 1$, $|Y_0(x)| \leq 1$, we conclude in the following problem: Find a function $H^{(0)}(x,0)$, regular for $x \in G_{\mathcal{M}}$, and continuous for $x \in \mathcal{M} \cup G_{\mathcal{M}}$:

$$Re(iH^{(0)}(x,0)) = Re\left(\frac{-iC(x,Y_0(x))(1 - \rho^*)}{A(x,Y_0(x))}\right) = w(x), \ x \in \mathcal{M}. \tag{18}$$

The problem defined in (18) is transformed into a Dirichlet problem on the unit circle $\mathcal{C}$, using the mappings $x = f_0(z) : G_{\mathcal{C}} \to G_{\mathcal{M}}$, $z = f(x) : G_{\mathcal{M}} \to G_{\mathcal{C}}$. Then, the condition (18) becomes $Re(i\tilde{G}(z)) = w(f_0(z))$, $z \in \mathcal{C}$, where $\tilde{G}(z) = H^{(0)}(f_0(z))$ regular in $G_{\mathcal{C}}$, and continuous in $\mathcal{C} \cup G_{\mathcal{C}}$. Its unique solution is

$$H^{(0)}(x,0) = \frac{-1}{2\pi} \int_{|t|=1} \frac{w(f_0(t))(t + f(x))}{t - f(x)} \frac{dt}{t} + N, \ x \in \mathcal{M} \cup G_{\mathcal{M}}, \tag{19}$$

where N some constant, determined by substituting $x = 0 \in G_{\mathcal{M}}$ in (19), and having in mind that $H^{(0)}(0,0) = 1 - \rho^*$. Using similar arguments, we can determine $H^{(0)}(0,y)$. Following the lines in Subsect. 4.1, we can finally obtain $H^{(n)}(x,y)$, $n = 0, 1, 2$.

In order everything to be well defined we have to compute the conformal mapping and its inverse; see [2]. Firstly, we parametrize $\mathcal{M}$ in polar coordinates, i.e., $\mathcal{M} = \{x : x = \rho(\phi)\exp(i\phi), \phi \in [0, 2\pi]\}$. Following [2], $\rho(\phi) = \frac{\delta(\phi)}{\cos(\phi)}$, where $\delta(\phi)$, is the unique solution of $\delta - \cos(\phi)\sqrt{m(\delta)}$, $\phi \in [0, 2\pi]$. Then,

$$f_0(z) = z\exp[\frac{1}{2\pi}\int_0^{2\pi}\log\{\rho(\psi(\omega))\}\frac{e^{i\omega}+z}{e^{i\omega}-z}d\omega], \ |z| < 1,$$
$$\psi(\phi) = \phi - \int_0^{2\pi}\log\{\rho(\psi(\omega))\}\cot(\frac{\omega-\phi}{2})d\omega, \ 0 \le \phi \le 2\pi, \tag{20}$$

i.e., $\psi(.)$ is uniquely obtained as the solution of Theodorsen integral equation with $\psi(\phi) = 2\pi - \psi(2\pi - \phi)$, where $f_0(0) = 0$ and $f_0(z) = \overline{f_0(\bar{z})}$.

Performance Measures: Denote by $H_1^{(n)}(x,y)$, $H_2^{(n)}(x,y)$, the derivatives of $H^{(n)}(x,y)$, $n = 0, 1, 2$, with respect to x and y, respectively. Then, $E(Q_k) = H_k^{(0)}(1,1) + H_k^{(1)}(1,1) + H_k^{(2)}(1,1)$, $k = 1, 2$. Due to space constraints we only focus on the case $\mu_1\mu_2 \ne d_1d_2$ and the derivation of $E(Q_1)$. Using (2), (5), (16),

$$H_1^{(0)}(1,1) = \frac{\lambda_1(d_1+\mu_2)+\mu_2(\theta t_1+\mu qr_1)-d_1(\mu(p+qr_2+\theta(t_0+t_2)))}{M_x}H_1^{(0)}(1,0)$$
$$+ \frac{[\lambda_1(d_1+\mu_2)+\mu_2(\theta t_1+\mu qr_1)]H^{(0)}(1,0)+[\lambda_1(d_2+\mu_1)+d_2(\theta t_1+\mu qr_1)]H^{(0)}(1,0)}{M_x}$$
$$- \frac{\lambda_1(d_1+d_2)H^{(0)}(0,0)+(\tilde{\lambda}_1+\mu_2(\theta t_1+\mu qr_1))H^{(0)}(1,1)}{M_x},$$
$$H_1^{(1)}(1,1) = \frac{\alpha H_1^{(0)}(1,1)-(d_1+\mu_2)H_1^{(0)}(1,0)-\mu qr_1-\theta t_1}{\mu+\theta},$$
$$H_1^{(2)}(1,1) = \frac{\theta}{\nu}(H_1^{(1)}(1,1)+t_1 H^{(0)}(1,1)),$$
$$H_1^{(0)}(1,0) = \frac{\mu_1\mu_2^*(\frac{x_*}{x_*-1})^{r_1}(\frac{x^*}{x^*-1})^{r_2}}{\mu_1\mu_2-d_1d_2}\exp[\frac{1}{2i\pi}\int_{|t|=1}\frac{\log\{J(t)\}f(1)}{t(t-f(1))}dt]$$
$$\times\{\frac{r_1}{x_*-1}+\frac{r_2}{x^*-1}+\frac{1}{2i\pi}\int_{|t|=1}\frac{\log\{J(t)\}f'(1)}{(t-f(1))^2}dt\}. \tag{21}$$

$E(Q_1)$ is obtained using (9), (17), (21). Similarly, we can obtain $E(Q_2)$.

5   The Symmetrical System

We now focus on the symmetrical system assuming hereon that $\lambda_j = \bar{\lambda}$, $\mu_j = \bar{\mu}$, $\mu_j^* = \mu^*$, $j = 1, 2$, $r_1 = r_2 = 1/2$, $t_1 = t_2 = t$. Thus, $\alpha = 2(\bar{\lambda}+\bar{\mu})$, $d_1 = d_2 := d = \bar{\mu} - \mu^*$, $\tilde{\lambda}_1 = \tilde{\lambda}_2 = \tilde{\lambda}$. This model is of a great interest since we are able to obtain closed form expressions for the expected number of packets stored at each relay node without solving a boundary value problem.

Note that, $H_1^{(0)}(1,1) = H_2^{(0)}(1,1)$, $H^{(0)}(1,0) = H^{(0)}(0,1)$ and $H_1^{(0)}(1,0) = H_2^{(0)}(0,1)$. Let $E(Q_j^{(n)}) = H_j^{(n)}(1,1)$, $j = 1, 2$, $n = 0, 1, 2$. Using (5), (8), (10),

$$E(Q_1^{(0)}) = \frac{(2\bar{\lambda}+\theta+\frac{\mu q}{2})[\bar{\lambda}+\frac{\mu q}{2}+\bar{\lambda}H^{(0)}(1,1)]+H_1^{(0)}(1,0)[d(\mu p+\theta t_0)+\mu^*(\frac{\mu q}{2}+\theta t)-\bar{\lambda}(d+\bar{\mu})]}{\bar{\mu}(\mu p+\theta t_0)-\bar{\lambda}}, \tag{22}$$

where $\bar{\mu}(\mu p + \theta t_0) > \tilde{\lambda}$ (i.e., the stability condition). Setting $y = x$ in (5),

$$\frac{d}{dx}[H^{(0)}(x,x)]|_{x=1} = \frac{2\bar{\lambda}(\bar{\lambda}+\frac{\mu q}{2})+\bar{\lambda}(2\bar{\lambda}+\theta t+\frac{\mu q}{2})H^{(0)}(1,1)+(d+\bar{\mu})(\mu p+\theta t_0-2\bar{\lambda})H_1^{(0)}(1,1)}{2(\bar{\mu}(\mu p+\theta t_0)-\tilde{\lambda})}, \tag{23}$$

Clearly, due to the symmetry, $\frac{d}{dx}[H^{(0)}(x,x)]|_{x=1} = H_1^{(0)}(1,1) + H_2^{(0)}(1,1) = 2H_1^{(0)}(1,1)$, and thus after some algebra, we can obtain from (22), (23),

$$E(Q_1^{(0)}) = \frac{1}{\mu^*(\mu+\theta)(\bar{\mu}(\mu p+\theta t_0)-\bar{\lambda})} \{\bar{\lambda}(2\bar{\lambda}+\theta t+\frac{\mu q}{2})H^{(0)}(1,1)[\bar{\mu}(\mu p+\theta t_0)$$
$$+\mu^*(\frac{\mu q}{2}+\theta t) - \bar{\lambda}(d+\bar{\mu})] + (\bar{\lambda}+\frac{\mu q}{2})[2\bar{\lambda}\bar{\mu}(\mu+\theta)+(d+\bar{\mu})(\mu p+\theta t_0)(\frac{\mu q}{2}+\theta t)]\}.$$

Using (2), (3), (22) we can easily obtain after simple calculations,

$$E(Q_1^{(1)}) = \frac{(\mu+\theta)[\alpha E(Q_1^{(0)})-(d+\bar{\mu})H_1^{(0)}(1,0)]-(2\theta t+\mu q)(\bar{\lambda}+\frac{\mu q}{2}+\bar{\lambda}H^{(0)}(1,1))}{(\mu+\theta)^2}, \qquad (24)$$
$$E(Q_1^{(2)}) = \frac{\theta}{\nu}E(Q_1^{(1)}),$$

where $H_1^{(0)}(1,0)$ is found by (22) and $\tilde{\lambda} = \bar{\lambda}(2\bar{\lambda}+2\bar{\mu}+\theta t+\mu\frac{q}{2})$. Therefore, $E(Q_1) = E(Q_2) = E(Q_1^{(0)}) + E(Q_1^{(1)})(1+\frac{\theta}{\nu})$.

6   A Special Case: A Traditional Cognitive Network

Let $\mu_i = \phi_i\mu_i^*$, $i = 1, 2$, $\phi_1 + \phi_2 = 1$. In the following we focus on a special case of the model by giving absolute priority to R_1. In such a case $\phi_1 = 1$, and thus, $A(x,y) = 0$. Thus, when both relay nodes are non-empty, R_2 remains silent (i.e., R_1 is the primary user and is allowed to access the spectrum at any time, and R_2 is the secondary user and has to transmit opportunistically by taking advantage of the idle periods of the primary nodes; [18]). Then, (5) is now written as

$$\tilde{R}(x,y)H^{(0)}(x,y) = \tilde{B}(x,y)H^{(0)}(0,y) + \tilde{C}(x,y)(1-\rho^*), \qquad (25)$$

where,

$$\tilde{R}(x,y) = xy(\tilde{\lambda}_2(y-1)+\tilde{\lambda}_1(x-1)) + \mu_1^*y[(\mu(p+qr_2)$$
$$+\theta(t_0+t_2))(1-x) + (\theta t_2+\mu qr_2)(y-1)],$$
$$\tilde{B}(x,y) = \mu_2^*x[(y-1)(\mu(p+qr_1)+\theta(t_0+t_1)-\lambda_2 y) + (1-x)(\theta t_1+\mu qr_1+\lambda_1 y)]$$
$$+\mu_1^*y[(1-x)(\mu(p+qr_2)+\theta(t_0+t_2)-\lambda_1 x) + (y-1)(\theta t_2+\mu qr_2+\lambda_2 x)],$$
$$\tilde{C}(x,y) = \mu_2^*x[(1-y)(\mu(p+qr_1)+\theta(t_0+t_1)-\lambda_2 y) + (x-1)(\theta t_1+\mu qr_1+\lambda_1 y)].$$

Using Rouche's theorem we can easily prove that the kernel $\tilde{R}(x,y)$ has a unique zero, say $x = \xi(y)$, inside the unit circle. For such a x, the right hand side of (25) vanishes, and we obtain

$$H^{(0)}(0,y) = \frac{-\tilde{C}(\xi(y),y)(1-\rho^*)}{\tilde{B}(\xi(y),y)},$$

Substituting back in (25), and using (2), (3), we obtain,

$$H^{(0)}(x,y) = \frac{(1-\rho^*)[\tilde{C}(x,y)\tilde{B}(\xi(y),y)-\tilde{C}(\xi(y),y)\tilde{B}(x,y)]}{\tilde{B}(\xi(y),y)\tilde{R}(x,y)},$$
$$H^{(1)}(x,y) = \frac{(1-\rho^*)}{\tilde{B}(\xi(y),y)\tilde{R}(x,y)T(x,y)}\{\tilde{C}(\xi(y),y)[R(x,y)(\mu_1^*-\mu_2^*)$$
$$-(\lambda+\mu_1^*)\tilde{B}(x,y)] + \tilde{B}(\xi(y),y)[(\lambda+\mu_1^*)\tilde{C}(x,y) - \mu_1^*R(x,y)]\},$$
$$H^{(2)}(x,y) = \frac{\theta[1+t_1(x-1)+t_2(y-1)]}{\nu}H^{(1)}(x,y),$$

where, $T(x,y) = \mu+\theta + (\theta t_1+\mu qr_1)(x-1) + (\theta t_2+\mu qr_2)(y-1)$.

7   Numerical Examples

Numerical Example 1: The case $\mu_1\mu_2 \neq d_1d_2$ (Symmetrical model) We focus on
the symmetrical model studied in Sect. 5. Set $\theta = 0.7$, $t_0 = 0.2$, $t = 0.4$, $p = 0.7$.
In Fig. 2 we observe how $E(Q_1)(= E(Q_2))$ evolves for increasing values of $\bar{\lambda}$ and
μ^*. In particular, in Fig. 2 (left) we can observe the effect of the service rate
at node D on $E(Q_1)$. As expected, by increasing the packet generation rate $\bar{\lambda}$,
$E(Q_1)$ increases. That increase becomes more apparent when μ decreases. Similar results can be obtained from Fig. 2 (right), where the effect of recovery rate
ν is depicted. In both figures, we have to mention the benefits of retransmission
control, in presence of μ^* (recall that μ^* is the retransmission rate of a relay node
when the other is empty). In Fig. 3 we observe $E(Q_1)$ as a function of $(\bar{\mu}, \mu^*)$
(left, $\mu = 5$), and (μ, μ^*) (right, $\bar{\mu} = 4$). In the former case, we can observe how
sensitive is $E(Q_1)$, as we slightly increase $\bar{\lambda}$, and especially when μ^* takes small
values ("pessimistic" retransmission control).

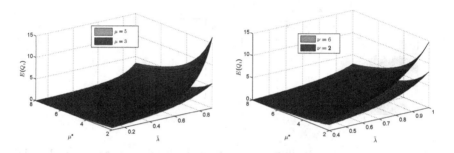

Fig. 2. Effect of μ (for $\nu = 6$, left), and ν (for $\mu = 5$, left) on $E(Q_1)$.

In the latter case, we can observe the impact of service rate μ on $E(Q_1)$.
Clearly, if we slightly increase $\bar{\lambda}$ from 1 to 1.2, $E(Q_1)$ increases dramatically as
μ decreases. That increase seems to be irreversible even when we increase μ^*.

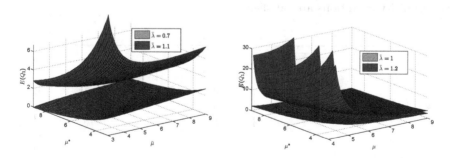

Fig. 3. $E(Q_1)$ as a function of $(\bar{\mu}, \mu^*)$ (left, $\mu = 5$), and (μ, μ^*) (right, $\bar{\mu} = 4$).

Numerical Example 2: The case $\mu_1\mu_2 \neq d_1d_2$ *(Stability region)* Fig. 4 (left) provides set of arrival rate vectors $(\widetilde{\lambda}_1, \widetilde{\lambda}_2)$ for which the system is stable, for $(\mu_1, \mu_2) = (4, 5)$, $\mu_1^* = 6$. Recall that $\widetilde{\lambda}_j$, $j = 1, 2$, is the rate at which packets flow into R_j (see Sect. 3). We can observe that when we increase μ_1^* from 6 (Blue+Red region) to 12 (Blue+Yellow region) the stability region changes significantly, since the increase in μ_1^* allows R_1 to retransmits faster. As a result, R_1 can handle more packets, and thus, we can allow larger values for $\widetilde{\lambda}_1$.

Numerical Example 3: The case $\mu_1\mu_2 \neq d_1d_2$ *(Simulations)* In the following we perform simulations experiments that show that measures Γ_1, Γ_2 introduced in Subsect. 3.1 allow to delimit accurate stability/instability regions. More precisely, we study the dynamics of the relay nodes depending on the values of Γ_1, Γ_2. Set $\lambda_2 = 4$, $\mu = 10$, $\mu_1 = 10$, $\mu_2 = 12$, $\mu_1^* = 15$, $\theta = 3$, $p = 0.8$, $r_1 = r_2 = 0.5$, $t_0 = 0.4$, $t_1 = t_2 = 0.3$, $\nu = 5$. Relay node dynamics when $\lambda_1 = 3$, $\mu_2 = 10$ are presented in Fig. 4 (right).

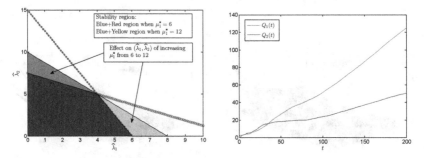

Fig. 4. Effect of μ_1^* on the stability region (left) and relay dynamics for $\lambda_1 = 3$ (right). (Color figure online)

There, we observe that the system becomes unstable, since the conditions of Theorem 1 are violated, i.e., $M_x = 76.3 > 0$, $M_y = 17.4 > 0$, $\Gamma_1 > 0$, $\Gamma_2 > 0$. In Fig. 5 we let $\lambda_1 = 2$ and in the left hand-side sub-figure we have $M_x = -28.4 < 0$, $M_y = 5.4 > 0$, and $\Gamma_1 < 0$, $\Gamma_2 < 0$. It is easily seen that the system is stable, as the ergodicity conditions are satisfied.

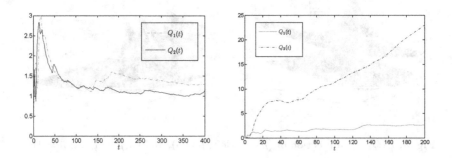

Fig. 5. Relay dynamics for $\mu_2^* = 18$ (left) and $\mu_2^* = 5$ (right).

In Fig. 5 (right), we observe the impact of the proposed queue-aware (cognitive) protocol. In particular, when we decrease μ_2^* from 18 to 5, ergodicity conditions are violated since $\Gamma_2 > 0$. As a result R_2 becomes unstable.

Acknowledgment. The author is grateful to the PC chairs and the anonymous referees for the valuable remarks, from which the presentation of the paper has benefited. He would also like to thank Dr. N. Pappas (Linköping University, Sweden), and Dr. T. Phung-Duc (University of Tsukuba, Japan) for their valuable comments.

References

1. Avrachenkov, K., Nain, P., Yechiali, U.: A retrial system with two input streams and two orbit queues. Queueing Syst. **77**, 1–31 (2014)
2. Cohen, J.W., Boxma, O.: Boundary Value Problems in Queueing Systems Analysis. North Holland Publishing Company, Amsterdam (1983)
3. Bing, B.: Emerging Technologies in Wireless LANs: Theory, Design, and Deployment. Cambridge University Press, New York (2007)
4. Bonald, T., Borst, S., Hegde, N., Proutiere, A.: Wireless data performance in multi-cell scenarios, In: Proceedings of ACM Sigmetrics/Performance 2004, pp. 378–388. ACM, New York (2004)
5. Borst, S., Jonckheere, M., Leskela, L.: Stability of parallel queueing systems with coupled service rates. Discrete Event Dyn. Syst. **18**(4), 447–472 (2008)
6. Cover, M., Gamal, A.: Capacity theorems for the relay channel. IEEE Trans. Infor. Theory **25**(5), 572–584 (1979)
7. Dimitriou, I.: A queueing model with two types of retrial customers and paired services. Ann. Oper. Res. **238**(1), 123–143 (2016)
8. Dimitriou, I.: A two class retrial system with coupled orbit queues. Probab. Eng. Inf. Sci. **31**(2), 139–179 (2017)
9. Dimitriou, I.: A queueing system for modeling cooperative wireless networks with coupled relay nodes and synchronized packet arrivals. Perform. Eval. (2017). doi:10.1016/j.peva.2017.04.002
10. Dimitriou, I.: A retrial queue to model a two-relay cooperative wireless system with simultaneous packet reception. In: Wittevrongel, S., Phung-Duc, T. (eds.) ASMTA 2016. LNCS, vol. 9845, pp. 123–139. Springer, Cham (2016). doi:10.1007/978-3-319-43904-4_9
11. Fayolle, G., Iasnogorodski, R., Malyshev, V.: Random Walks in the Quarter-Plane, Algebraic Methods, Boundary Value Problems and Applications. Springer-Verlag, Berlin (2017)
12. Gakhov, F.D.: Boundary Value Problems. Pergamon Press, Oxford (1966)
13. Haykin, S.: Cognitive radio: brain-empowered wireless communications. IEEE J. Sel. Areas Commun. **23**, 201–220 (2005)
14. Liu, K., Sadek, A., Su, W., Kwasinski, A.: Cooperative Communications and Networking. Cambridge University Press, Cambridge (2008)
15. Mitola, J., Maguire, G.: Cognitive radio: making software radios more personal. IEEE Pers. Commun. **6**(4), 13–18 (1999)
16. Nosratinia, A., Hunter, T.E., Hedayat, A.: Cooperative communication in wireless networks. Comm. Mag. **42**(10), 74–80 (2004)
17. Pappas, N., Kountouris, M., Ephremides, A., Traganitis, A.: Relay-assisted multiple access with full-duplex multi-packet reception. IEEE Trans. Wirel. Commun. **14**, 3544–3558 (2015)

18. Sadek, A., Liu, K., Ephremides, A.: Cognitive multiple access via cooperation: protocol design and performance analysis. IEEE Trans. Inf. Theory **53**(10), 3677–3696 (2007)
19. Sendonaris, A., Erkip, E., Aazhang, B.: User cooperation diversity-Part I: system description. IEEE Trans. Commun. **51**, 1927–1938 (2003)
20. Resing, J., Ormeci, L.: A tandem queueing model with coupled processors. Oper. Res. Lett. **31**, 383–389 (2003)
21. Van Leeuwaarden, J., Resing, J.: A tandem queue with coupled processors: computational issues. Queueing Syst. **50**, 29–52 (2005)

Stability and Delay Analysis of an Adaptive Channel-Aware Random Access Wireless Network

Ioannis Dimitriou[1]([⊠]) and Nikolaos Pappas[2]

[1] Department of Mathematics, University of Patras, 26500 Patras, Greece
idimit@math.upatras.gr
[2] Department of Science and Technology, Linköping University,
60174 Norrköping, Sweden
nikolaos.pappas@liu.se

Abstract. In this work, we consider an asymmetric two-user random access wireless network with interacting nodes, time-varying links and multipacket reception capabilities. The users are equipped with infinite capacity buffers where they store arriving packets that will be transmitted to a destination node. Moreover, each user employs a general transmission control protocol under which, it adapts its transmission probability based both on the state of the other user, and on the channel state information according to a Gilbert-Elliot model. We study a two-dimensional discrete time Markov chain, investigate its stability condition, and show that its steady state performance is expressed in terms of a solution of a Riemann-Hilbert boundary value problem. Moreover, for the symmetrical system, we provide closed form expressions for the average delay at each user node. Numerical results are obtained and show insights in the system performance.

Keywords: Boundary value problem · Random access · Multipacket reception · Adaptive transmission · Channel aware · Stability region · Delay analysis · Gilbert-Elliott channel

1 Introduction

Random access has re-gained attention recently because of the need for massive uncoordinated access in large networks which will be common in the fifth generation of mobile networks (5G) era [1,4,28] (not an exhaustive list). Thus, the study of random access in large networks is of major importance [4]. However, there are still many unanswered fundamental questions regarding the performance of random access even in small networks [14,18].

When the traffic in a network is bursty, a meaningful performance measure is the stable throughput region i.e. the stability region, which gives the set of arrival rates such that there exist transmission probabilities under which the system is stable [24,32,34]. Characterizing the stability region in random access networks

© Springer International Publishing AG 2017
N. Thomas and M. Forshaw (Eds.): ASMTA 2017, LNCS 10378, pp. 63–80, 2017.
DOI: 10.1007/978-3-319-61428-1_5

is a well known difficult problem because of the interaction of the queues. The stability region is a throughput metric with bounded delay guarantees, but in most of the works appeared in the past, stability and delay were studied in isolation. The stability region of a two-user random access network with traditional collision channel has been studied in [32,34,35]. A more detailed treatment of stable throughput for various cases can be found in [22]. In [21], the stability region of a cognitive radio system of two source-destination pairs in the presence of imperfect sensing was studied. For a three-user random access network with collision channel model the stability region was obtained in [34], while for the case of more than three users the exact stability region is not known yet except for some derived bounds given in [24].

Although stable throughput region in random access systems has been studied for several cases, the delay performance is so far overlooked in the research. 5G was proposed aiming to enhance the networking capabilities of mobile users [1,28]. Differentiated from 4G, benefits offered by 5G will be much more than the increased maximum throughput [1]. Thus, the rapid growth on supporting real-time applications requires delay-based guarantees. However, the characterization of delay even in small networks with random access is rather difficult, even for the traditional collision model [26]. Although the traditional collision channel model is suitable for wire-line communications, it is not an appropriate model for probabilistic reception in wireless multiple access scenarios. Moreover, most of the related works are based on the strong assumption of the absolute symmetry of the network; e.g., [15,27,33]. More importantly they did not take into account the impact of time-varying links, e.g., [27,33], as well as the ability of a node to adapt its transmission probability based on the knowledge of the status of neighbor nodes, which in turn, leads to self-aware networks. Note that this feature is very common in cognitive radios [5,6,21,25].

Contribution. In this work, we study an asymmetric two-user random access wireless network where the user's transmission probability is adapted based both on the status of the other user, and on the channel state. We model the state of the wireless channels as a Gilbert-Elliot model that changes between a "good" and a "bad" state. Our motivation stems from the fact that the channel conditions may vary, and thus, the success probability of a packet transmission is affected. Moreover, we take account advances in multiuser detection, which allow the receiver to employ multipacket reception (MPR) capabilities, and to correctly receive at most one user packet, even if many users transmit (i.e., the "capture" effect).

We analyze a system of two queues, we investigate the stable throughput, and the queueing delay. Finally, we evaluate numerically the derived analytical results. Our system is modeled as a two-dimensional discrete time Markov chain, and we show that its steady-state performance is expressed in terms of the solution of a Riemann-Hilbert problem [19]. For related works on queueing systems using the theory of boundary value problems see e.g. [2,3,8–13,16,17,26,31,36]. To the best of our knowledge there is no other work in the related literature in which exact expressions for the stability conditions of a random access system where a user adapts its transmission probability based on its "knowledge" about both the status of the other, and of the channel state. In such a case, we take into account the wireless interference as well as the complex interdependence among users' nodes due to the shared medium. Clearly, such a protocol leads to substantial performance gains, since each user exploits the idle slots of the other. More importantly, besides its applicability, our work is also theoretically oriented, since we provide, for the first time, an exact detailed analysis of an asymmetric adapted random access wireless system with MPR capabilities, and obtain the generating function of the stationary joint queue length distribution with the aid of the theory of boundary value problems.

The rest of the paper is organized as follows. In Sect. 2 we describe the model in detail, and derive the fundamental functional equation. In Sect. 3 we obtain some important results for the following analysis, and investigate the stability conditions. Section 4 is devoted to the formulation and solution of two boundary value problems, the solution of which provides the generating function of the joint queue length distribution of user nodes. In Sect. 5 we obtain explicit expressions for the average delay at each user for the symmetrical system. Finally, in Sect. 6 we obtain useful numerical examples that show insights in the system performance.

2   Model Description and the Functional Equation

We consider an asymmetric random access system consisting of $N = 2$ users communicating with a common receiver. Each user has an infinite capacity buffer, in which stores arriving and backlogged packets. Packets have equal length and the time is divided into slots corresponding to the transmission time of a packet. At the beginning of each slot, there is an opportunity for the user node k, $k = 1, 2$, to transmit a packet to the receiver.

The channel of a particular link is independent between users, and varies between slots according to a Gilbert-Elliott model, where it can be in one of two states at any given time slot: the good state, denoted by "G" and the bad state, denoted by "B". The channel state is assumed to be fixed during a slot duration and varies in an independent and identically distributed (i.i.d.) manner between slots [1]. The long term proportion of time in which user k's channel is in state i is denoted by $s_i^{(k)}$, $i \in \{B, G\}$, $k \in \{1, 2\}$; and can be obtained either through channel measurements or through a physical model of the channels.

Users have perfect channel knowledge and adjust their transmission probabilities (transmission control) according to the channel state. Due to the interference among the stations we consider the following opportunistic policy: If both stations are non empty, station k, $k = 1, 2$, transmits a packet according to a Bernoulli stream with probability q_{ik} independently, $\bar{q}_{ik}$ is the probability that

---

[1] This model can capture the case where the wireless channel has strong interference by another external network or when the channel is in deep fading. In both cases we can assume that the channel is in the bad state. It is outside of the scope of this version of the paper to consider detailed physical layer considerations.

station k does not make a transmission in a slot, given that his channel is in state $i \in \{B, G\}$. If station 1 (resp. 2) is the only non-empty, it changes its transmission probability to q_{ik}^* independently[2], $\bar{q}_{ik}^* = 1 - q_{ik}^*$ is the probability that station k does not make a transmission in the given slot. Note that in our case, a node is aware of the state of the other node. This is a common assumption in the literature related to cognitive wireless networks [5,6,22,25].

The success of a transmission depends on the underlying channel model. The MPR channel model used in this paper is a generalized form of the packet erasure model. In particular we focus on a subclass of MPR model, the "capture" channels [7,23,37]. In such a case, at most one packet can be successfully received at the destination if more than one nodes transmit. A common assumption in wireless networks is that a packet can be decoded correctly by the receiver if the received SINR (Signal-to-Interference-plus-Noise-Ratio) exceeds a certain threshold. The set of transmitting nodes in a given timeslot is denoted by T. Let $f_{ik/T}$ the probability that a packet transmitted from node k with channel state i is successfully decoded at the destination, i.e., $f_{ik/T} = Pr(\gamma_{ik/T} > \theta)$, where $\gamma_{ik/T}$ denotes the SINR of the signal transmitted from node i with channel state k at the receiver given the channel states of the transmitters and the threshold for the successful decoding θ, which depends on the modulation scheme, target bit error rate and the number of bits in the packet. Without loss of generality we assume that when the channel state is "bad" and transmission fails with probability 1,[3] i.e., $f_{B,k/T} = 0$, $k = 1, 2$. Furthermore, let $\tilde{f}_{i,k/\{i,k\}}$ be the success probability of node k when it is the only non empty ($\tilde{f}_{B,k/\{B,k\}} = 0$, $k = 1, 2$). We consider the following success probabilities for nodes 1 and 2

$$\tilde{f}_{G,1/\{G,1\}} > f_{G,1/\{G,1\}} > f_{G,1/\{G,1;B,2\}} > f_{G,1/\{G,1;G,2\}},$$
$$\tilde{f}_{G,2/\{G,2\}} > f_{G,2/\{G,2\}} > f_{G,2/\{B,1;G,2\}} > f_{G,2/\{G,1;G,2\}}.$$

Note that the success probability when a packet is transmitted in the presence of interference cannot exceed the success probability when it is transmitted alone. Let also denote by $f_{0/\{G,1;G,2\}} = 1 - f_{G,1/\{G,1;G,2\}} - f_{G,2/\{G,1;G,2\}}$, $f_{0/\{G,k\}} = 1 - f_{G,k/\{G,k\}}$, $\tilde{f}_{0/\{G,k\}} = 1 - \tilde{f}_{G,k/\{G,k\}}$, the probabilities that no packets will be successfully transmitted.

In case of unsuccessful transmissions the packets have to be re-transmitted in a later slot. We assume that the receiver gives an instantaneous (error-free) feedback of all the packets that were successful in a slot at the end of the slot to all the nodes. The nodes remove the successfully transmitted packets from their buffers while unsuccessful packets are retained.

Let $\{A_{k,n}\}_{n \in \mathbf{N}}$ be a sequence of i.i.d. random variables where $A_{k,n}$ represents the number of packets which arrive at buffer k in the interval $(n, n + 1]$, with $E(A_{k,n}) = \hat{\lambda}_k < \infty$. Denote by $D(x, y) = \lim_{n \to \infty} E(x^{A_{1,n}} y^{A_{2,n}})$, $|x| \leq 1$,

---

[2] We consider the general case for q_{ik}^*, this can handle cases where the node cannot transmit with probability one even if it is transmitting alone. Such a scenario may occur when the nodes are subject to energy limitations. The study of energy harvesting in random access networks has been considered in [5,20,29,30].

[3] We assume this mostly for simplicity, however, our work can be extended for the case that the success probability is not zero when the channel is in the bad state.

$|y| \leq 1$, the generating function of the stationary joint distribution of the number of arriving packets in any slot. In this work we assume that the arrival processes at both user nodes are independent and geometrically distributed, i.e.,

$$D(x,y) = [(1 + \widehat{\lambda}_1(1 - x))(1 + \widehat{\lambda}_2(1 - y))]^{-1}.$$

Denote by $N_{k,n}$ the number of packets at user node k at the beginning of the n-th slot. Then, $Y_n = (N_{1,n}, N_{2,n})$ is a discrete time Markov chain with state space $E = \{(i,j) : i,j = 0,1,2,\ldots\}$. The users' queues evolve as

$$N_{k,n+1} = [N_{k,n} - D_{k,n}]^+ + A_{k,n}, \; k = 1,2, \tag{1}$$

where $D_{k,n}$ is the number of departures from user k queue at time slot n. Let $H(x,y)$ be the generating function of the joint stationary queue process, viz.

$$H(x,y) = \lim_{n \to \infty} E(x^{N_{1,n}} y^{N_{2,n}}), \; |x| \leq 1, \; |y| \leq 1.$$

Then, by exploiting (1) (see Appendix A), we obtain after lengthy calculations,

$$R(x,y)H(x,y) = A(x,y)H(x,0) + B(x,y)H(0,y) + C(x,y)H(0,0), \tag{2}$$

where,

$$R(x,y) = D^{-1}(x,y) + s_G^{(1)} q_{G1}\widehat{q}_{12}(1 - \tfrac{1}{x}) + s_G^{(2)} q_{G2}\widehat{q}_{21}(1 - \tfrac{1}{y}),$$
$$A(x,y) = s_G^{(2)} q_{G2}\widehat{q}_{21}(1 - \tfrac{1}{y}) + d_{1,2}(1 - \tfrac{1}{x}),$$
$$B(x,y) = s_G^{(1)} q_{G1}\widehat{q}_{12}(1 - \tfrac{1}{x}) + d_{2,1}(1 - \tfrac{1}{y}),$$
$$C(x,y) = d_{2,1}(\tfrac{1}{y} - 1) + d_{1,2}(\tfrac{1}{x} - 1),$$

and,

$$\widehat{q}_{km} = (s_G^{(m)}\bar{q}_{Gm} + s_B^{(m)}\bar{q}_{Bm})f_{G,k/G,k} + s_B^{(m)}q_{Bm}f_{G,k/\{G,k;B,m\}}$$
$$+ s_G^{(m)}q_{Gm}f_{G,k/\{G,k;G,m\}}, \; k,m \in \{1,2\}, \; k \neq m,$$
$$d_{k,m} = s_G^{(k)}q_{Gk}\widehat{q}_{km} - s_G^{(k)}q_{Gk}^*\tilde{f}_{G,k/G,k}, \; k,m \in \{1,2\}, \; k \neq m.$$

Some interesting relations can be obtained directly from the functional Eq. (2). Taking $y = 1$, dividing by $x - 1$ and taking $x \to 1$ in (2), and vice versa, yield the following "conservation of flow" relations:

$$\widehat{\lambda}_1 = s_G^{(1)} q_{G1}\widehat{q}_{12}(1 - H(0,1)) - d_{1,2}(H(1,0) - H(0,0)),$$
$$\widehat{\lambda}_2 = s_G^{(2)} q_{G2}\widehat{q}_{21}(1 - H(1,0)) - d_{2,1}(H(0,1) - H(0,0)). \tag{3}$$

Using (3), we distinguish the analysis in two cases, which differ both from the modeling and the technical point of view:

1. For $\dfrac{q_{G1}\widehat{q}_{12}}{q_{G1}^* f_{G,1/G,1}} + \dfrac{q_{G2}\widehat{q}_{21}}{q_{G2}^* f_{G,2/G,2}} = 1$, Eq. (3) yields

$$H(0,0) = 1 - \left(\frac{\widehat{\lambda}_1}{s_G^{(1)} q_{G1}^* \tilde{f}_{G,1/G,1}} + \frac{\widehat{\lambda}_2}{s_G^{(2)} q_{G2}^* \tilde{f}_{G,2/G,2}} \right) = 1 - \rho.$$

2. For $\frac{q_{G1}\widehat{q}_{12}}{q_{G1}^* f_{G,1/G,1}} + \frac{q_{G2}\widehat{q}_{21}}{q_{G2}^* f_{G,2/G,2}} \neq 1$, Eq. (3) yields

$$
\begin{aligned}
H(1,0) &= \frac{d_{2,1}\widehat{\lambda}_1 + s_G^{(1)} q_{G1}\widehat{q}_{12}(s_G^{(2)} q_{G2}^* \tilde{f}_{G2/\{G2\}} - \widehat{\lambda}_2) + d_{2,1}s_G^{(1)} q_{G1}^* \tilde{f}_{G,1/\{G,1\}} H(0,0)}{s_G^{(1)} q_{G1}\widehat{q}_{12}s_G^{(2)} q_{G2}\widehat{q}_{21} - d_{1,2}d_{2,1}}, \\
H(0,1) &= \frac{d_{1,2}\widehat{\lambda}_2 + s_G^{(2)} q_{G2}\widehat{q}_{21}(s_G^{(1)} q_{G1}^* \tilde{f}_{G1/\{G1\}} - \widehat{\lambda}_1) + d_{1,2}s_G^{(2)} q_{G2}^* \tilde{f}_{G,2/\{G,2\}} H(0,0)}{s_G^{(1)} q_{G1}\widehat{q}_{12}s_G^{(2)} q_{G2}\widehat{q}_{21} - d_{1,2}d_{2,1}}.
\end{aligned}
\tag{4}
$$

3 Preparatory Analysis

We now focus on the derivation of some preparatory results in view of the resolution of the functional Eq. (2). We first investigate the stability criteria, and then, we focus on the analysis of the kernel equation $R(x,y) = 0$.

3.1 Stability Region

Based on the concept of stochastic dominant systems [32,34], we derive the stability region, i.e., the set of vectors $(\widehat{\lambda}_1, \widehat{\lambda}_2)$, for which our system is stable.

Lemma 1. *The stability region $\mathcal{R}$ for a fixed transmission probability vector $\mathbf{q} := [q_{G1}, q_{G2}, q_{G1}^*, q_{G2}^*]$ is given by*

1. *In case $\frac{q_{G1}\widehat{q}_{12}}{q_{G1}^* f_{G,1/G,1}} + \frac{q_{G2}\widehat{q}_{21}}{q_{G2}^* f_{G,2/G,2}} \neq 1$, $\mathcal{R} = \mathcal{R}_1 \cup \mathcal{R}_2$ where,*

$$
\begin{aligned}
\mathcal{R}_1 &= \{(\widehat{\lambda}_1, \widehat{\lambda}_2) : \widehat{\lambda}_1 < s_G^{(1)} q_{G1}^* \tilde{f}_{G,1/\{G,1\}} + d_{1,2}\frac{\widehat{\lambda}_2}{s_G^{(2)} q_{G2}\widehat{q}_{21}}, \widehat{\lambda}_2 < s_G^{(2)} q_{G2}\widehat{q}_{21}\}, \\
\mathcal{R}_2 &= \{(\widehat{\lambda}_1, \widehat{\lambda}_2) : \widehat{\lambda}_2 < s_G^{(2)} q_{G2}^* \tilde{f}_{G,2/\{G,2\}} + d_{2,1}\frac{\widehat{\lambda}_1}{s_G^{(1)} q_{G1}\widehat{q}_{12}}, \widehat{\lambda}_1 < s_G^{(1)} q_{G1}\widehat{q}_{12}\}.
\end{aligned}
$$

2. *In case $\frac{q_{G1}\widehat{q}_{12}}{q_{G1}^* f_{G,1/G,1}} + \frac{q_{G2}\widehat{q}_{21}}{q_{G2}^* f_{G,2/G,2}} = 1$, $\mathcal{R} = \{(\widehat{\lambda}_1, \widehat{\lambda}_2) : \rho < 1\}$.*

Proof: In order to determine the stability region, we apply the stochastic dominance technique developed in [32,34], which consists of considering hypothetical auxiliary systems that closely parallel the operation of the original system but dominate it in a well defined manner. Under this approach, we consider the R_1, and R_2 dominant systems. In the R_k dominant system, whenever the queue of user node k, $k = 1,2$ empties, it continues to transmit "dummy" packets.

The dominant system has the following properties [32]: (i) the queue lengths in the dominant system are no shorter than the queues in the original system. Thus, if the queues in the dominant system are stable, then, the queues in the original system are stable as well, (ii) the two systems coincide at saturation, that is, if the queue of user 1 never empties (that is, if it is saturated or unstable), then the dominant system, and the original system are indistinguishable; and thus, the instability of the dominant system implies the instability of the original system. Clearly, (i) and (ii) imply that the stability condition of the dominant system is a necessary and sufficient for the stability of the original system and hence, the stable throughput regions of both systems coincide for fixed transmission probabilities.

Thus, in R_1, user node 1 never empties, and its service rate depends on whether user node 2 is empty or not. On the other hand, user node 2 "sees" a constant service rate. Therefore, in the R_1 dominant system, $\widehat{\lambda}_2 < s_G^{(2)} q_{G2} \widehat{q}_{21}$. Moreover, the stability condition for the user node 1 is given by,

$$\widehat{\lambda}_1 < s_G^{(1)} q_{G1}^* \tilde{f}_{G,1/\{G,1\}} \left(1 - \frac{\widehat{\lambda}_2}{s_G^{(2)} q_{G2} \widehat{q}_{21}}\right) + s_G^{(1)} q_{G1} \widehat{q}_{12} \frac{\widehat{\lambda}_2}{s_G^{(2)} q_{G2} \widehat{q}_{21}}.$$

Thus, the sufficient condition for the ergodicity of the R_1 dominant system is,

$$\widehat{\lambda}_1 < s_G^{(1)} q_{G1}^* \tilde{f}_{G,1/\{G,1\}} + \frac{d_{1,2} \widehat{\lambda}_2}{s_G^{(2)} q_{G2} \widehat{q}_{21}}, \text{ and } \widehat{\lambda}_2 < s_G^{(2)} q_{G2} \widehat{q}_{21}. \tag{5}$$

Similarly, the sufficient ergodicity condition of the R_2 system is given by,

$$\widehat{\lambda}_1 < s_G^{(1)} q_{G1} \widehat{q}_{12}, \text{ and } \widehat{\lambda}_2 < s_G^{(2)} q_{G2}^* \tilde{f}_{G,2/\{G,2\}} + \frac{d_{2,1} \widehat{\lambda}_1}{s_G^{(1)} q_{G1} \widehat{q}_{12}}. \tag{6}$$

Combining the sufficient conditions for both the dominant systems (i.e., (5), (6)) yields the sufficiency part of the lemma. The necessary part of the lemma follows by an "indistinguishability" argument similar to the one used in [32]. □

Remark: $\mathcal{R}$ is a convex polyhedron when $\frac{q_{G1} \widehat{q}_{12}}{q_{G1}^* f_{G,1/G,1}} + \frac{q_{G2} \widehat{q}_{21}}{q_{G2}^* f_{G,2/G,2}} \geq 1$. When equality holds, the region is a triangle and coincides with the case of time-sharing. Convexity is an important property since it corresponds to the case when parallel concurrent transmissions are preferable to time-sharing.

3.2   Analysis of the Kernel

We now provide some detailed properties of the kernel $R(x, y)$, which are important for the formulation and solution of the boundary value problems. Clearly,

$$R(x, y) = a(x)y^2 + b(x)y + c(x) = \widehat{a}(y)x^2 + \widehat{b}(y)x + \widehat{c}(y),$$

where, $a(x) = \widehat{\lambda}_2 x (\widehat{\lambda}_1(x - 1) - 1)$, $c(x) = -s_G^{(2)} q_{G2} \widehat{q}_{21} x$, $b(x) = x(\widehat{\lambda} + \widehat{\lambda}_1 \widehat{\lambda}_2 + s_G^{(1)} q_{G1} \widehat{q}_{12} + s_G^{(2)} q_{G2} \widehat{q}_{21}) - s_G^{(1)} q_{G1} \widehat{q}_{12} - \widehat{\lambda}_1 (1 + \widehat{\lambda}_2) x^2$, $\widehat{a}(y) = \widehat{\lambda}_1 y (\widehat{\lambda}_2(y - 1) - 1)$, $\widehat{c}(y) = -s_G^{(1)} q_{G1} \widehat{q}_{12} y$, $\widehat{b}(y) = y(\widehat{\lambda} + \widehat{\lambda}_1 \widehat{\lambda}_2 + s_G^{(1)} q_{G1} \widehat{q}_{12} + s_G^{(2)} q_{G2} \widehat{q}_{21}) - s_G^{(2)} q_{G2} \widehat{q}_{21} - \widehat{\lambda}_2 (1 + \widehat{\lambda}_1) y^2$. The roots of $R(x, y) = 0$ are $X_\pm(y) = \frac{-\widehat{b}(y) \pm \sqrt{D_y(y)}}{2\widehat{a}(y)}$, $Y_\pm(x) = \frac{-b(x) \pm \sqrt{D_x(x)}}{2a(x)}$, where $D_y(y) = \widehat{b}(y)^2 - 4\widehat{a}(y)\widehat{c}(y)$, $D_x(x) = b(x)^2 - 4a(x)c(x)$.

Lemma 2. *For* $|y| = 1$, $y \neq 1$, *the kernel equation* $R(x, y) = 0$ *has exactly one root* $x = X_0(y)$ *such that* $|X_0(y)| < 1$. *For* $\widehat{\lambda}_1 < s_G^{(1)} q_{G1} \widehat{q}_{12}$, $X_0(1) = 1$. *Similarly, we can prove that* $R(x, y) = 0$ *has exactly one root* $y = Y_0(x)$, *such that* $|Y_0(x)| \leq 1$, *for* $|x| = 1$.

Proof: It is easily seen that $R(x,y) = \frac{xy - \Psi(x,y)}{xyD(x,y)}$, where $\Psi(x,y) = D(x,y)[xy - y(x-1)s_G^{(1)}q_{G1}\widehat{q}_{12} - x(y-1)s_G^{(2)}q_{G2}\widehat{q}_{21}]$, where for $|x| \leq 1$, $|y| \leq 1$, $\Psi(x,y)$ is a generating function of a proper probability distribution. Now, for $|y| = 1$, $y \neq 1$ and $|x| = 1$ it is clear that $|\Psi(x,y)| < 1 = |xy|$. Thus, from Rouché's theorem, $xy - \Psi(x,y)$ has exactly one zero inside the unit circle. Therefore, $R(x,y) = 0$ has exactly one root $x = X_0(y)$, such that $|x| < 1$. For $y = 1$, $R(x,1) = 0$ implies $(x-1)[\widehat{\lambda}_1 - \frac{s_G^{(1)}q_{G1}\widehat{q}_{12}}{x}] = 0$. Therefore, for $y = 1$, and since $\widehat{\lambda}_1 < s_G^{(1)}q_{G1}\widehat{q}_{12}$, the only root of $R(x,1) = 0$ for $|x| \leq 1$, is $x = 1$.   □

Lemma 3. *The algebraic function* $Y(x)$, *defined by* $R(x, Y(x)) = 0$, *has four real branch points* $0 < x_1 < x_2 \leq 1 < x_3 < x_4 < \frac{1+\widehat{\lambda}_1}{\lambda_1}$. *Moreover,* $D_x(x) < 0$, $x \in (x_1, x_2) \cup (x_3, x_4)$ *and* $D_x(x) > 0$, $x \in (-\infty, x_1) \cup (x_2, x_3) \cup (x_4, \infty)$. *Similarly,* $X(y)$, *defined by* $R(X(y), y) = 0$, *has four real branch points* $0 \leq y_1 < y_2 \leq 1 < y_3 < y_4 < \frac{1+\widehat{\lambda}_2}{\lambda_2}$, *and* $D_x(y) < 0$, $y \in (y_1, y_2) \cup (y_3, y_4) <$ *and* $D_x(y) > 0$, $y \in (-\infty, y_1) \cup (y_2, y_3) \cup (y_4, \infty)$.

Proof: The proof is based on simple algebraic arguments; see also [13].   □

Consider now the cut planes: $\widetilde{\mathbb{C}}_x = C_x - ([x_1, x_2] \cup [x_3, x_4])$, $\widetilde{\mathbb{C}}_y = C_y - ([y_1, y_2] \cup [y_3, y_4])$, where C_x, C_y the complex planes of x, y, respectively. In $\widetilde{\mathbb{C}}_x$ (resp. $\widetilde{\mathbb{C}}_y$), let $Y_0(x)$ (resp. $X_0(y)$) be the zero of $R(x, Y(x)) = 0$ (resp. $R(X(y), y) = 0$) with the smallest modulus.

Lemma 4. *1. For* $y \in [y_1, y_2]$, *the algebraic function* $X(y)$ *lies on a closed contour* $\mathcal{M}$, *which is symmetric with respect to the real line and defined by*

$$|x|^2 = m(Re(x)), \quad m(\delta) = \frac{s_G^{(1)}q_{G1}\widehat{q}_{12}}{\widehat{\lambda}_1(1+\widehat{\lambda}_2 - \widehat{\lambda}_2\zeta(\delta))}, \quad |x|^2 \leq \frac{s_G^{(1)}q_{G1}\widehat{q}_{12}}{\widehat{\lambda}_1(1+\widehat{\lambda}_2 - \widehat{\lambda}_2 y_2)},$$

where, $k(\delta) := \widehat{\lambda} + \widehat{\lambda}_1\widehat{\lambda}_2 + s_G^{(1)}q_{G1}\widehat{q}_{12} + s_G^{(2)}q_{G2}\widehat{q}_{21} - 2\widehat{\lambda}_1(1+\widehat{\lambda}_2)\delta$ *and,* $\zeta(\delta) = \frac{k(\delta) - \sqrt{k^2(\delta) - 4s_G^{(2)}q_{G2}\widehat{q}_{21}\widehat{\lambda}_2(1+\widehat{\lambda}_1(1-2\delta))}}{2\widehat{\lambda}_2(1+\widehat{\lambda}_1(1-2\delta))}$.

Set $\beta_0 := \sqrt{\frac{s_G^{(1)}q_{G1}\widehat{q}_{12}}{\widehat{\lambda}_1(1+\widehat{\lambda}_2 - \widehat{\lambda}_2 y_2)}}$, $\beta_1 := -\sqrt{\frac{s_G^{(1)}q_{G1}\widehat{q}_{12}}{\widehat{\lambda}_1(1+\widehat{\lambda}_2 - \widehat{\lambda}_2 y_1)}}$ *the extreme right and left point of* $\mathcal{M}$, *respectively.*

2. For $x \in [x_1, x_2]$, *the algebraic function* $Y(x)$ *lies on a closed contour* $\mathcal{L}$, *which is symmetric with respect to the real line and defined by*

$$|y|^2 = v(Re(y)), \quad v(\delta) = \frac{s_G^{(2)}q_{G2}\widehat{q}_{21}}{\widehat{\lambda}_2(1+\widehat{\lambda}_1 - \widehat{\lambda}_1\theta(\delta))}, \quad |y|^2 \leq \frac{s_G^{(2)}q_{G2}\widehat{q}_{21}}{\widehat{\lambda}_2(1+\widehat{\lambda}_1 - \widehat{\lambda}_1 x_2)},$$

where $l(\delta) := \widehat{\lambda} + \widehat{\lambda}_1\widehat{\lambda}_2 + s_G^{(1)}q_{G1}\widehat{q}_{12} + s_G^{(2)}q_{G2}\widehat{q}_{21} - 2\widehat{\lambda}_2(1+\widehat{\lambda}_1)\delta$, *and* $\theta(\delta) = \frac{l(\delta) - \sqrt{l^2(\delta) - 4s_G^{(1)}q_{G1}\widehat{q}_{12}\widehat{\lambda}_1(1+\widehat{\lambda}_2(1-2\delta))}}{2\widehat{\lambda}_1(1+\widehat{\lambda}_2(1-2\delta))}$.

Set $\eta_0 := \sqrt{\frac{s_G^{(2)}q_{G2}\widehat{q}_{21}}{\widehat{\lambda}_2(1+\widehat{\lambda}_1 - \widehat{\lambda}_1 x_2)}}$, $\eta_1 := -\sqrt{\frac{s_G^{(2)}q_{G2}\widehat{q}_{21}}{\widehat{\lambda}_2(1+\widehat{\lambda}_1 - \widehat{\lambda}_1 x_1)}}$ *the extreme right and left point of* $\mathcal{L}$, *respectively.*

Proof: We only focus on the first part. For $y \in [y_1, y_2]$, $D_y(y) < 0$, so $X_{\pm}(y)$ are complex conjugates. Thus, $|X(y)|^2 = \frac{s_G^{(1)} q_{G1} \widehat{q}_{12}}{\widehat{\lambda}_1 (1 + \widehat{\lambda}_2 - \widehat{\lambda}_2 y)} = g(y)$. It also follows that

$$Re(X(y)) = \frac{y(\widehat{\lambda} + \widehat{\lambda}_1 \widehat{\lambda}_2 + s_G^{(1)} q_{G1} \widehat{q}_{12} + s_G^{(2)} q_{G2} \widehat{q}_{21}) - s_G^{(2)} q_{G2} \widehat{q}_{21} - \widehat{\lambda}_2 (1 + \widehat{\lambda}_1) y^2}{2 \widehat{\lambda}_1 y (1 + \widehat{\lambda}_2 - \widehat{\lambda}_2 y)}. \tag{7}$$

Clearly, $g(y)$ is an increasing function for $y \in [0, 1]$ and thus, $|X(y)|^2 \le g(y_2) = \beta_0$. Using simple algebraic considerations we can prove that, $X_0(y_1) = \beta_1$ is the extreme left point of $\mathcal{M}$. Finally, $\zeta(\delta)$ is derived by solving (7) for y with $\delta = Re(X(y))$, and taking the solution such that $y \in [0, 1]$. □

4 The Boundary Value Problems

In the following, we distinguish the analysis in two cases, which differ from both the modeling and the technical point of view.

4.1 A Dirichlet Boundary Value Problem

Assume that $\frac{q_{G1} \widehat{q}_{12}}{q_{G1}^* f_{G,1/G,1}} + \frac{q_{G2} \widehat{q}_{21}}{q_{G2}^* f_{G,2/G,2}} = 1$. Then, $A(x, y) = \frac{s_G^{(2)} q_{G2} \widehat{q}_{21}}{d_{2,1}} B(x, y)$.
Therefore, for $y \in \mathcal{D}_y = \{y \in \mathcal{C}_y : |y| \le 1, |X_0(y)| \le 1\}$,

$$s_G^{(2)} q_{G2} \widehat{q}_{21} H(X_0(y), 0) + d_{2,1} H(0, y) + \frac{s_G^{(2)} q_{G2} \widehat{q}_{21} C(X_0(y), y)}{A(X_0(y), y)}(1 - \rho) = 0. \tag{8}$$

For $y \in \mathcal{D}_y - [y_1, y_2]$ both $H(X_0(y), 0)$, $H(0, y)$ are analytic and the right-hand side in (8) can be analytically continued up to the slit $[y_1, y_2]$, or equivalently, for $x \in \mathcal{M}$

$$s_G^{(2)} q_{G2} \widehat{q}_{21} H(x, 0) + d_{2,1} H(0, Y_0(x)) + \frac{s_G^{(2)} q_{G2} \widehat{q}_{21} C(x, Y_0(x))}{A(x, Y_0(x))}(1 - \rho) = 0. \tag{9}$$

Then, by multiplying both sides of (9) by the imaginary complex number i, and noticing that $H(0, Y_0(x))$ is real for $x \in \mathcal{M}$, since $Y_0(x) \in [y_1, y_2]$, we have

$$Re(iH(x, 0)) = Re\left(-i\frac{C(x, Y_0(x))}{A(x, Y_0(x))}\right)(1 - \rho), x \in \mathcal{M}. \tag{10}$$

To proceed, we have to check for possible poles of $H(x, 0)$ in $S_x := G_{\mathcal{M}} \cap \bar{D}_x^c$, where $G_{\mathcal{U}}$ be the interior domain bounded by $\mathcal{U}$, and $D_x = \{x : |x| < 1\}$, $\bar{D}_x = \{x : |x| \le 1\}$, $\bar{D}_x^c = \{x : |x| > 1\}$. These poles, if exist, they coincide with the zeros of $A(x, Y_0(x))$ in S_x (see Appendix B). In order to solve (10) we must firstly conformally transform the problem from $\mathcal{M}$ to the unit circle $\mathcal{C}$. Let the conformal mapping $z = \gamma(x) : G_{\mathcal{M}} \to G_{\mathcal{C}}$, and its inverse given by $x = \gamma_0(z) : G_{\mathcal{C}} \to G_{\mathcal{M}}$. Then, we have the following problem: Find a function $\tilde{T}(z) = H^{(0)}(\gamma_0(z))$ regular for $z \in G_{\mathcal{C}}$, and continuous for $z \in \mathcal{C} \cup G_{\mathcal{C}}$ such that, $Re(i\tilde{G}(z)) = w(\gamma_0(z))$, $z \in \mathcal{C}$.

In the following, we need a representation of $\mathcal{M}$ in polar coordinates, i.e., $\mathcal{M} = \{x : x = \rho(\phi) \exp(i\phi), \phi \in [0, 2\pi]\}$. In the following we summarize the basic steps; see [8]. Since $0 \in G_{\mathcal{M}}$, for each $x \in \mathcal{M}$, a relation between its absolute value and its real part is given by $|x|^2 = m(Re(x))$ (see Lemma 4). Given the angle ϕ of some point on $\mathcal{M}$, the real part of this point, say $\delta(\phi)$, is the solution of $\delta - \cos(\phi)\sqrt{m(\delta)}$, $\phi \in [0, 2\pi]$. Since $\mathcal{M}$ is a smooth, egg-shaped contour, the solution is unique. Clearly, $\rho(\phi) = \frac{\delta(\phi)}{\cos(\phi)}$, and the parametrization of $\mathcal{M}$ in polar coordinates is fully specified. Then, the mapping from $z \in G_{\mathcal{C}}$ to $x \in G_{\mathcal{M}}$, where $z = e^{i\phi}$ and $x = \rho(\psi(\phi))e^{i\psi(\phi)}$, satisfying $\gamma_0(0) = 0$ and $\gamma_0(z) = \overline{\gamma_0(\bar{z})}$ is uniquely determined by (see [8]),

$$
\begin{aligned}
\gamma_0(z) &= z \exp[\tfrac{1}{2\pi} \int_0^{2\pi} \log\{\rho(\psi(\omega))\} \tfrac{e^{i\omega}+z}{e^{i\omega}-z} d\omega], \ |z| < 1, \\
\psi(\phi) &= \phi - \int_0^{2\pi} \log\{\rho(\psi(\omega))\} \cot(\tfrac{\omega-\phi}{2}) d\omega, \ 0 \le \phi \le 2\pi,
\end{aligned}
\tag{11}
$$

i.e., the angular deformation $\psi(.)$ is uniquely determined as the solution of Theodorsen integral equation with $\psi(\phi) = 2\pi - \psi(2\pi - \phi)$. If $H(x,0)$ has no poles in S_x, the solution of the problem defined in (10) is,

$$
H(x,0) = -\tfrac{1-\rho}{2\pi} \int_{|t|=1} f(t) \tfrac{t+\gamma(x)}{t-\gamma(x)} \tfrac{dt}{t} + S, \ x \in \mathcal{M},
\tag{12}
$$

where $f(t) = Re\left(-i\frac{C(\gamma_0(t),Y_0(\gamma_0(t)))}{A(\gamma_0(t),Y_0(\gamma_0(t)))}\right)$. S is a constant to be defined by setting $x = 0 \in G_{\mathcal{M}}$ in (12), and using the fact that $H(0,0) = 1 - \rho$, $\gamma(0) = 0$ (In case $H(x,0)$ has a pole, i.e. $x = \bar{x}$, we have still a Dirichlet problem for the function $(x - \bar{x})H(x,0)$; see Appendix B). Following the discussion above, $S = (1-\rho)(1 + \tfrac{1}{2\pi} \int_{|t|=1} f(t) \tfrac{dt}{t})$. Setting $t = e^{i\phi}$, $\gamma_0(e^{i\phi}) = \rho(\psi(\phi))e^{i\psi(\phi)}$, we arrive after some algebra in,

$$
f(e^{i\phi}) = \frac{d_{1,2}s_G^{(2)}q_{G2}^*\tilde{f}_{G,1/\{G,1\}}\sin(\psi(\phi))(1-Y_0(\gamma_0(e^{i\phi}))^{-1})}{\rho(\psi(\phi))\{[s_G^{(2)}q_{G2}\hat{q}_{21}(1-Y_0^{-1}(\gamma_0(e^{i\phi})))+d_{1,2}(1-\frac{\cos(\psi(\phi))}{\rho(\psi(\phi))})]^2+(d_{1,2}\frac{\sin(\psi(\phi))}{\rho(\psi(\phi))})^2\}},
$$

which is an odd function of ϕ. Thus, $S = 1 - \rho$. Substituting back in (12):

$$
H(x,0) = (1-\rho)\{1 + \tfrac{2\gamma(x)i}{\pi} \int_0^\pi \tfrac{f(e^{i\phi})\sin(\phi)d\phi}{1-2\gamma(x)\cos(\phi)-\gamma(x)^2}\}, \ x \in G_{\mathcal{M}}.
\tag{13}
$$

Similarly, we can determine $H(0,y)$ by solving another Dirichlet boundary value problem on the contour $\mathcal{L}$. Then, using (2) we uniquely obtain $H(x,y)$.

4.2 A Homogeneous Riemann-Hilbert Boundary Value Problem

In case $\frac{q_{G1}\hat{q}_{12}}{q_{G1}^*\tilde{f}_{G,1/G,1}} + \frac{q_{G2}\hat{q}_{21}}{q_{G2}^*\tilde{f}_{G,2/G,2}} \ne 1$, consider the following transformation:

$$
\begin{aligned}
G(x) &:= H(x,0) + \frac{s_G^{(1)}q_{G1}^*\tilde{f}_{G,1/\{G,1\}}d_{2,1}H(0,0)}{d_{1,2}d_{2,1}-s_G^{(1)}q_{G1}\hat{q}_{12}s_G^{(2)}q_{G2}\hat{q}_{21}}, \\
L(y) &:= H(0,y) + \frac{s_G^{(2)}q_{G2}^*\tilde{f}_{G,2/\{G,2\}}d_{1,2}H(0,0)}{d_{1,2}d_{2,1}-s_G^{(1)}q_{G1}\hat{q}_{12}s_G^{(2)}q_{G2}\hat{q}_{21}}.
\end{aligned}
$$

Then, for $y \in D_y$,

$$
A(X_0(y),y)G(X_0(y)) = -B(X_0(y),y)L(y).
\tag{14}
$$

For $y \in D_y - [y_1, y_2]$ both $G(X_0(y))$, $L(y)$ are analytic and the right-hand side in (14) can be analytically continued up to the slit $[y_1, y_2]$, or equivalently,

$$A(x, Y_0(x))G(x) = -B(x, Y_0(x))L(Y_0(x)), \ x \in \mathcal{M}. \tag{15}$$

Clearly, $G(x)$ is holomorphic in D_x, and continuous in $\bar{D}_x$. However, $G(x)$ might have poles in $S_x = G_\mathcal{M} \cap \bar{D}_x^c$. These poles (if exist) coincide with the zeros of $A(x, Y_0(x))$ in S_x; see Appendix B. For $y \in [y_1, y_2]$, let $X_0(y) = x \in \mathcal{M}$, and realize that $Y_0(X_0(y)) = y$ (note that $B(x, Y_0(x)) \neq 0$, $x \in \mathcal{M}$). Taking into account the poles of $G(x)$, and noticing that $L(Y_0(x))$ is real for $x \in \mathcal{M}$,

$$Re[iU(x)\tilde{G}(x)] = 0, \ x \in \mathcal{M},$$
$$U(x) = \frac{A(x, Y_0(x))}{(x-\bar{x})^r B(x, Y_0(x))}, \ \tilde{G}(x) = (x - \bar{x})^r G(x), \tag{16}$$

where $r = 0, 1$, whether $\bar{x}$ is zero or not of $A(x, Y_0(x))$ in S_x (see Appendix B). Thus, $\tilde{G}(x)$ is regular for $x \in G_\mathcal{M}$, continuous for $x \in \mathcal{M} \cup G_\mathcal{M}$, and $U(x)$ is a non-vanishing function on $\mathcal{M}$. As usual, we must firstly conformally transform the problem (16) from $\mathcal{M}$ to the unit circle $\mathcal{C}$. Let the conformal mapping $z = \gamma(x) : G_\mathcal{M} \rightarrow G_\mathcal{C}$, and its inverse given by $x = \gamma_0(z) : G_\mathcal{C} \rightarrow G_\mathcal{M}$.

Then, the Riemann-Hilbert problem formulated in (16) is reduced to the following: Find a function $F(z) := \tilde{H}(\gamma_0(z))$, regular in $G_\mathcal{C}$, continuous in $G_\mathcal{C} \cup \mathcal{C}$ such that, $Re[iU(\gamma_0(z))F(z)] = 0, \ z \in \mathcal{C}$.

To proceed with the solution of the boundary value problem we have to determine its index $\chi = \frac{-1}{\pi}[arg\{U(x)\}]_{x \in \mathcal{M}}$, where $[arg\{U(x)\}]_{x \in \mathcal{M}}$, denotes the variation of the argument of the function $U(x)$ as x moves along $\mathcal{M}$ in the positive direction, provided that $U(x) \neq 0$, $x \in \mathcal{M}$. Following [16] we have,

Lemma 5. *1. If $\widehat{\lambda}_2 < s_G^{(2)} q_{G2} \widehat{q}_{21}$, then $\chi = 0$ is equivalent to*

$$\frac{dA(x, Y_0(x))}{dx}\Big|_{x=1} < 0 \Leftrightarrow \widehat{\lambda}_1 < s_G^{(1)} q_{G1}^* \tilde{f}_{G,1/\{G,1\}} + d_{1,2}\frac{\widehat{\lambda}_2}{s_G^{(2)} q_{G2} \widehat{q}_{21}},$$
$$\frac{dB(X_0(y), y)}{dy}\Big|_{y=1} < 0 \Leftrightarrow \widehat{\lambda}_2 < s_G^{(2)} q_{G2}^* \tilde{f}_{G,2/\{G,2\}} + d_{2,1}\frac{\widehat{\lambda}_1}{s_G^{(1)} q_{G1} \widehat{q}_{12}}.$$

2. If $\widehat{\lambda}_2 \geq s_G^{(2)} q_{G2} \widehat{q}_{21}$, $\chi = 0$ is equivalent to $\frac{dB(X_0(y), y)}{dy}\Big|_{y=1} < 0 \Leftrightarrow \widehat{\lambda}_2 <$ $s_G^{(2)} q_{G2}^* \tilde{f}_{G,2/\{G,2\}} + d_{2,1}\frac{\widehat{\lambda}_1}{s_G^{(1)} q_{G1} \widehat{q}_{12}}.$

Thus, under stability conditions (see Lemma 1), the problem defined in (16) has a unique solution for $x \in G_\mathcal{M}$ given by,

$$H(x, 0) = K(x - \bar{x})^r e^{[\frac{1}{2i\pi} \int_{|t|=1} \frac{\log\{J(t)\}dt}{t-\gamma(x)}]} - \frac{s_G^{(1)} q_{G1}^* \tilde{f}_{G,1/\{G,1\}} d_{2,1} H(0,0)}{d_{1,2} d_{2,1} - s_G^{(1)} q_{G1} \widehat{q}_{12} s_G^{(2)} q_{G2} \widehat{q}_{21}}, \tag{17}$$

where K is a constant, $J(t) = \frac{\overline{U_1(t)}}{U_1(t)}$, $U_1(t) = U(\gamma_0(t))$. Setting $x = 0$ in (17) we derive a relation among K, $H(0, 0)$. Now set $x = 1 \in G_\mathcal{M}$ in (17), and use the first in (4) to obtain K, $H(0, 0)$. Substituting back in (17) we finally obtain,

$$H(x,0) = \frac{\widehat{\lambda}_1 d_{2,1} + s_G^{(1)} q_{G1}\widehat{q}_{12}(s_G^{(2)} q_{G2}^* \tilde{f}_{G2/\{G2\}} - \widehat{\lambda}_2)}{(s_G^{(1)} q_{G1}\widehat{q}_{12}s_G^{(2)} q_{G2}\widehat{q}_{21} - d_{1,2}d_{2,1})(\bar{x}-1)^r} ((\bar{x} - x)^r$$

$$\times \exp[\frac{\gamma(x)-\gamma(1)}{2\pi i} \int_{|t|=1} \frac{\log\{J(t)\}}{(t-\gamma(x))(t-\gamma(1))} dt] \tag{18}$$

$$+ \frac{q_{G1}^* \tilde{f}_{G1/\{G1\}} d_{2,1}\bar{x}^r}{q_{G1}\widehat{q}_{12}s_G^{(2)} q_{G2}^* \tilde{f}_{G2/\{G2\}}} \exp[\frac{-\gamma(1)}{2\pi i} \int_{|t|=1} \frac{\log\{J(t)\}}{t(t-\gamma(1))} dt]), \; x \in G_{\mathcal{M}}.$$

Similarly, we can determine $H(0,y)$ by solving another Riemann-Hilbert boundary value problem on the closed contour $\mathcal{L}$. Then, using the fundamental functional Eq. (2) we uniquely obtain $H(x,y)$.

Performance Metrics: In the following we derive formulas for the expected number of packets, and the average delay at each user node in steady state, say M_i and D_i, $i = 1,2$, respectively. Denote by $H_1(x,y)$, $H_2(x,y)$ the derivatives of $H(x,y)$ with respect to x and y, respectively. Then, $M_i = H_i(1,1)$, and using Little's law $D_i = H_i(1,1)/\widehat{\lambda}_i$, $i = 1,2$. Using (2), (3) after simple calculations we have

$$M_1 = \frac{\widehat{\lambda}_1 + d_{1,2} H_1(1,0)}{s_G^{(1)} q_{G1}\widehat{q}_{12}}, \quad M_2 = \frac{\widehat{\lambda}_2 + d_{2,1} H_2(0,1)}{s_G^{(2)} q_{G2}\widehat{q}_{21}}. \tag{19}$$

We only focus on M_1, D_1 (similarly we can obtain M_2, D_2). Note that $H_1(1,0)$ can be obtained using (18) or (13) depending on the value of $\frac{q_{G1}\widehat{q}_{12}}{q_{G1}^* f_{G,1/G,1}} + \frac{q_{G2}\widehat{q}_{21}}{q_{G2}^* f_{G,2/G,2}}$. For $\frac{q_{G1}\widehat{q}_{12}}{q_{G1}^* f_{G,1/G,1}} + \frac{q_{G2}\widehat{q}_{21}}{q_{G2}^* f_{G,2/G,2}} \neq 1$, and using (18) we obtain,

$$H_1(1,0) = \frac{\widehat{\lambda}_1 d_{2,1} + s_G^{(1)} q_{G1}\widehat{q}_{12}(s_G^{(2)} q_{G2}^* \tilde{f}_{G1/\{G1\}} - \widehat{\lambda}_2)}{s_G^{(1)} q_{G1}\widehat{q}_{12}s_G^{(2)} q_{G2}\widehat{q}_{21} - d_{1,2}d_{2,1}} \{\frac{\gamma'(1)}{2\pi i} \int_{|t|=1} \frac{\log\{J(t)\}}{(t-\gamma(1))^2} dt$$

$$+ \frac{r}{1-\bar{x}} 1_{\{r=1\}} \}. \tag{20}$$

Substituting in (19) we obtain M_1, and dividing with $\widehat{\lambda}_1$, the average delay D_1. Note that the calculation of (11) requires the evaluation of integrals (11), and $\gamma(1)$, $\gamma'(1)$. For an efficient numerical procedure see [8], Section 4.1.

5   Explicit Expressions for the Symmetrical Model

In this section we consider the symmetrical model and obtain exact expressions for the average delay without computing the generating function of the stationary joint relay queue length distribution. As a symmetrical, we mean the model where $q_{Gk}^* = q_G^*$, $q_{ik} = q_i$, $i \in \{B,G\}$, $\lambda_k = \lambda$, $f_{i,k/\{i,k\}} = f_{i/\{i\}}$, $f_{G,k/\{G,k;G,m\}} = f_{G/\{G;G\}}$, $f_{G,k/\{G,k;B,m\}} = f_{G/\{G;B\}}$, $\tilde{f}_{G,k/\{G,k\}} = \tilde{f}_G$, $s_i^{(k)} = s_i$, $i \in \{G,B\}$, $k = 1,2$. Then, $\widehat{q}_{12} = \widehat{q}_{21} = \widehat{q}$ and $d_{1,2} = d_{2,1} = d$.

Due to the symmetry of the model, $H_1(1,1) = H_2(1,1)$, $H_1(1,0) = H_2(0,1)$. Note that $M_j = H_j(1,1)$, $j = 1,2$. Thus, using (2) we obtain,

$$M_1 = \frac{\widehat{\lambda} + dH_1(1,0)}{s_G q_G \widehat{q} - \widehat{\lambda}}, \tag{21}$$

where $s_G q_G \widehat{q} > \widehat{\lambda}$, due to the stability condition. Set $x = y$ in (2), differentiate it with respect to x at $x = 1$, and use the first in (3) to obtain,

$$M_1 + M_2 = 2M_1 = \frac{2\widehat{\lambda}-\widehat{\lambda}^2+2H_1(1,0)(s_G q_G \widehat{q}+d)}{s_G q_G \widehat{q}-\widehat{\lambda}}. \tag{22}$$

Using (21), (22), and applying Little's law we finally derive,

$$M_1 = M_2 = \frac{\widehat{\lambda}[2s_G q_G \widehat{q}+\widehat{\lambda}d]}{2s_G q_G^* f_G(s_G q_G \widehat{q}-\widehat{\lambda})}, \quad D_1 = D_2 = \frac{2s_G q_G \widehat{q}+\widehat{\lambda}d}{2s_G q_G^* f_G(s_G q_G \widehat{q}-\widehat{\lambda})}. \tag{23}$$

6   Numerical Results

Example 1: The symmetrical model. In this example, we focus on the symmetrical model and investigate the effect of transmission control on the average delay. We assume that $\widehat{f}_G = 0.9$, $f_{G/\{G\}} = 0.8$, $f_{G/\{G,B\}} = 0.7$, $f_{G/\{G,G\}} = 0.6$, $q_B = 0.5$. In Figs. 1 and 2 we also assume that $q_G^* = 0.9$. Recall that q_G^* is the transmission probability of a station when the other is empty.

In Fig. 1 (left) we observe that the average delay increases for increasing values of $\widehat{\lambda}$ by letting $q_G = 0.7$. More importantly, we can identify the advantage of transmission control regarding delay. Note that when the channel remains in the good state for longer period (i.e., $s_G = 0.9$), the users adapt their transmission probability q_G to 0.7, and thus, significant performance gains are achieved. Similar observations can be deduced from Fig. 1 (right), where we can see that the average delay decreases, as we increase q_G. That decrease becomes more apparent when the channel remains for longer period in the good state.

Figure 2 (left) shows the average delay as a function of $(\widehat{\lambda}, q_G)$. We can see how sensitive is the average delay when $\widehat{\lambda}$ increases, and especially, when the portion of time where the channel is in good state decreases. Similarly, when s_G decreases (see Fig. 2 (right)), the average delay increases rapidly, especially when q_G takes small values (this maybe happened when users falsely detect that the channel is in the bad state). Finally in Fig. 3 (left) we observe that when the channel is in good state for longer period (e.g., $s_G = 1$), the average delay decreases, even when q_G^* takes small values, a fact that justifies the importance of transmission control, from the delay point of view.

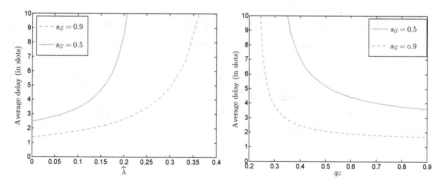

Fig. 1. Effect of transmission control on the average delay for $q_G = 0.7$ (left), and for $\widehat{\lambda} = 0.1$ (right).

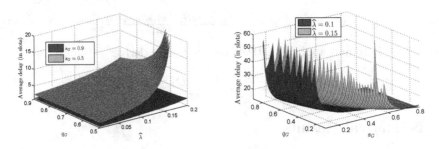

Fig. 2. Effect of transmission control on the average delay.

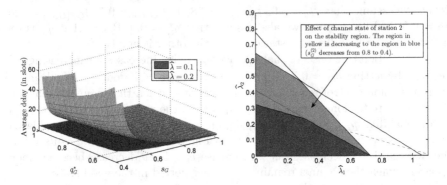

Fig. 3. Effect of transmission control (s_G, q_G^*) on the average delay for $q_G = 0.7$ (left), and effect of channel state on the stability region (right). (Color figure online)

Example 2: Stability region for the general model. In this example we focus on the general model, and specifically on the case $\frac{q_{G1}\widehat{q}_{12}}{q_{G1}^* f_{G,1/G,1}} + \frac{q_{G2}\widehat{q}_{21}}{q_{G2}^* f_{G,2/G,2}} \neq 1$. Our aim is to investigate the effect of channel state on the stability region. We assume that $s_G^{(1)} = 0.9$, $q_{G1} = 0.6$, $q_{G2} = 0.7$, $q_{B1} = 0.3$, $q_{B2} = 0.4$, $q_{G1}^* = 0.9 = q_{G2}^*$, and $\widehat{f}_{G,k/\{G,k\}} = 0.9$, $f_{G,k/\{G,k\}} = 0.8$, $f_{G,1/\{G,1;B,2\}} = f_{G,2/\{G,2;B,1\}} = 0.7$, $f_{G,k/\{G,1;G,2\}} = 0.6$, $k = 1, 2$.

In Fig. 3 (right) we observe the impact of channel state of user 2 on the stability region. In particular, when we decrease $s_G^{(2)}$ from 0.8 to 0.4, the stability region (i.e., the set of arrival vectors $(\widehat{\lambda}_1, \widehat{\lambda}_2)$ for which both queues are stable) apparently decreases. Note, that the adequate arrival rate referring to queue 2 is greatly reduced due to the change of $s_G^{(2)}$. This is expected, since when $s_G^{(2)} = 0.4$, the channel remains in the bad state for longer period and user 2 becomes reluctant to transmit. Thus, in order to ensure stability the $\widehat{\lambda}_2$ must be reduced.

7   Summary

In this work, we considered the problem of characterizing stability and delay behavior of an asymmetric adaptive two-user random access wireless network with MPR capabilities. Each user node use its knowledge about both the channel state (characterized according to the Gilbert-Elliot model), and the status of the other node, and accordingly adjusts its transmission probability. Stability conditions were investigated based on the concept of stochastic dominant systems. The generating function of the stationary joint queue length distribution was derived in terms of the solution of a Riemann-Hilbert boundary value problem. For the symmetrical system we also derived explicit expressions for the average queueing delay at each user node without solving a boundary value problem. Extensive numerical results shown insights into the system performance.

A Appendix

Regarding the derivation of (2), the queue evolution in (1) implies,

$$
\begin{aligned}
E(x^{N_{1,n+1}}y^{N_{2,n+1}}) &= D(x,y)\,(P(N_{1,n}=N_{2,n}=0)\\
&+E(x^{N_{1,n}}1_{\{N_{1,n}>0,N_{2,n}=0\}})[1+s_G^{(1)}q_{G1}^*\tilde{f}_{G,1/\{G,1\}}(\tfrac{1}{x}-1)]\\
&+E(y^{N_{2,n}}1_{\{N_{1,n}=0,N_{2,n}>0\}})[1+s_G^{(2)}q_{G2}^*\tilde{f}_{G,2/\{G,2\}}(\tfrac{1}{y}-1)\\
&+E(x^{N_{1,n}}y^{N_{2,n}}1_{\{N_{1,n}>0,N_{2,n}>0\}})[s_G^{(1)}q_{G1}(s_G^{(2)}\bar{q}_{G2}+s_B^{(2)}\bar{q}_{B2})\\
&\times(1+f_{G,1/\{G,1\}}(\tfrac{1}{x}-1))+s_G^{(2)}q_{G2}(s_G^{(1)}\bar{q}_{G1}+s_B^{(1)}\bar{q}_{B1})(1+f_{G,2/\{G,2\}}(\tfrac{1}{y}-1))]\\
&+s_G^{(1)}q_{G1}s_G^{(2)}q_{G2}(1+f_{G,1/\{G,1;G,2\}}(\tfrac{1}{x}-1)+f_{G,2/\{G,1;G,2\}}(\tfrac{1}{y}-1))\\
&+s_G^{(1)}q_{G1}s_B^{(2)}q_{B2}(1+f_{G,1/\{G,1;B,2\}}(\tfrac{1}{x}-1))+s_G^{(1)}q_{B1}s_G^{(2)}q_{G2}\\
&\times(1+f_{G,2/\{B,1;G,2\}}(\tfrac{1}{y}-1))+(s_G^{(1)}\bar{q}_{G1}+s_B^{(1)}\bar{q}_{B1})(s_G^{(2)}\bar{q}_{G2}+s_B^{(2)}\bar{q}_{B2})\\
&+s_B^{(1)}q_{B1}s_B^{(2)}q_{B2}(s_G^{(1)}\bar{q}_{G1}+s_B^{(1)}\bar{q}_{B1})+s_B^{(2)}q_{B2}(s_G^{(1)}\bar{q}_{G1}+s_B^{(1)}\bar{q}_{B1})+s_B^{(1)}\bar{q}_{B1}(s_G^{(2)}\bar{q}_{G2}+s_B^{(2)}\bar{q}_{B2})\Big),
\end{aligned}
$$

where $1_{\{A\}}$ denotes the indicator function of the event A. Note that

$$
\begin{aligned}
H(x,0)-H(0,0) &= \lim_{n\to\infty}E(x^{N_{1,n}}1_{\{N_{1,n}>0,N_{2,n}=0\}}),\\
H(0,y)-H(0,0) &= \lim_{n\to\infty}E(y^{N_{2,n}}1_{\{N_{1,n}=0,N_{2,n}>0\}}),\\
H(x,y)-H(x,0)-H(0,y)+H(0,0) &= \lim_{n\to\infty}E(x^{N_{1,n}}y^{N_{2,n}}1_{\{N_{1,n}>0,N_{2,n}>0\}}).
\end{aligned}
$$

B Appendix

In the following, we proceed with the study of the location of the intersection points of $R(x,y)=0$, $A(x,y)=0$ (resp. $B(x,y)$). These points (if exist) are potential singularities for the functions $H(x,0)$, $H(0,y)$, and thus, their investigation is crucial regarding the analytic continuation of $H(x,0)$, $H(0,y)$ outside the unit disk. We only focus on the intersection points of $R(x,y)=0$, $A(x,y)=0$.

For $x\in\tilde{C}_x$ and $R(x,y)=0$, $y=Y_\pm(x)$, the resultant in y of the two polynomials $R(x,y)$ and $A(x,y)$ is $Res_y(R,A;x)=x(x-1)s_G^{(2)}q_{G2}\hat{q}_{21}Z(x)$, where

$$
\begin{aligned}
Z(x) &= -\hat{\lambda}_1(s_G^{(2)}q_{G2}\hat{q}_{21}+(1+\hat{\lambda}_1)d_{1,2})x^2+x[(\hat{\lambda}+\hat{\lambda}_1\hat{\lambda}_2)d_{1,2}\\
&+(s_G^{(2)}q_{G2}\hat{q}_{21}+d_{1,2})s_G^{(1)}q_{G1}^*\tilde{f}_{G1/\{G1\}}]-s_G^{(1)}q_{G1}^*\tilde{f}_{G1/\{G1\}}d_{1,2}.
\end{aligned}
$$

Note also that $Z(0) > 0$ since $d_{1,2} < 0$, and $Z(1) > 0$, due to the stability conditions (see Lemma 1). If $q_{G1}^* \leq min\{1, \frac{s_G^{(2)} q_{G2} \hat{q}_{21} + (1+\hat{\lambda}_2) s_G^{(1)} q_{G1} \hat{q}_{12}}{(1+\hat{\lambda}_2) s_G^{(1)} \tilde{f}_{G1/\{G1\}}}\}$, then $\lim_{x \to \infty} Z(x) = -\infty$, and $Z(x) = 0$ has two roots of opposite sign, say $x_* < 0 < 1 < x^*$. If $\frac{s_G^{(2)} q_{G2} \hat{q}_{21} + (1+\hat{\lambda}_2) s_G^{(1)} q_{G1} \hat{q}_{12}}{(1+\hat{\lambda}_2) s_G^{(1)} \tilde{f}_{G1/\{G1\}}} < \alpha_1^* \leq 1$, then $\lim_{x \to \infty} Z(x) = +\infty$, and $Z(x) = 0$ has two positive roots, say $1 < \tilde{x}_* < x_3 < x_4 < \tilde{x}^*$ (due to the stability conditions). In the former case we have to check if $x^* \in S_x$, while in the latter case if $\tilde{x}_* \in S_x$. These zeros, if they lie in S_x such that $|Y_0(x)| \leq 1$, are poles of $A(x, y)$. Denote from hereon $\bar{x} = x^*$, if $\alpha_1^* \leq min\{1, \frac{s_G^{(2)} q_{G2} \hat{q}_{21} + (1+\hat{\lambda}_2) s_G^{(1)} q_{G1} \hat{q}_{12}}{(1+\hat{\lambda}_2) s_G^{(1)} \tilde{f}_{G1/\{G1\}}}\}$, and $\bar{x} = \tilde{x}_*$, if $\frac{s_G^{(2)} q_{G2} \hat{q}_{21} + (1+\hat{\lambda}_2) s_G^{(1)} q_{G1} \hat{q}_{12}}{(1+\hat{\lambda}_2) s_G^{(1)} \tilde{f}_{G1/\{G1\}}} < \alpha_1^* \leq 1$.

References

1. Alliance, N.: NGMN 5G white paper. Next generation mobile networks, White paper (2015)
2. Avrachenkov, K., Nain, P., Yechiali, U.: A retrial system with two input streams and two orbit queues. Queueing Syst. **77**, 1–31 (2014)
3. Boxma, O.: Two symmetric queues with alternating service and switching times. In: Gelenbe, E. (ed.) Performance 1984, pp. 409–431. North-Holland, Amsterdam (1984)
4. Chatzikokolakis, K., Kaloxylos, A., Spapis, P., Alonistioti, N., Zhou, C., Eichinger, J., Bulakci, Ö.: On the way to massive access in 5G: challenges and solutions for massive machine communications. In: Weichold, M., Hamdi, M., Shakir, M.Z., Abdallah, M., Karagiannidis, G.K., Ismail, M. (eds.) CrownCom 2015. LNICSSITE, vol. 156, pp. 708–717. Springer, Cham (2015). doi:10.1007/978-3-319-24540-9_58
5. Chen, Z., Pappas, N., Kountouris, M.: Energy harvesting in delay-aware cognitive shared access networks. In: IEEE ICC Workshops, Paris, France (2017)
6. Chen, Z., Pappas, N., Kountouris, M., Angelakis, V.: Throughput analysis of smart objects with delay constraints. In: IEEE 17th WoWMoM, Coimbra, Portugal (2016)
7. Cidon, H.K.I., Sidi, M.: Erasure, capture and random power level selection in multiple access systems. IEEE Trans. Commun. **36**, 263–271 (1988)
8. Cohen, J.W., Boxma, O.: Boundary Value Problems Queueing Systems Analysis. North Holland Publishing Company, Amsterdam (1983)
9. Dimitriou, I.: A two class retrial system with coupled orbit queues. Prob. Eng. Inf. Sci. **31**(2), 139–179 (2017)
10. Dimitriou, I.: A queueing model with two types of retrial customers and paired services. Ann. Oper. Res. **238**(1), 123–143 (2016)
11. Dimitriou, I.: A retrial queue to model a two-relay cooperative wireless system with simultaneous packet reception. In: Wittevrongel, S., Phung-Duc, T. (eds.) ASMTA 2016. LNCS, vol. 9845, pp. 123–139. Springer, Cham (2016). doi:10.1007/978-3-319-43904-4_9
12. Dimitriou, I.: A queueing system for modeling cooperative wireless networks with coupled relay nodes and synchronized packet arrivals. Perform. Eval. (2017). doi:10.1016/j.peva.2017.04.002

13. Dimitriou, I., Pappas, N.: Stable throughput and delay analysis of a random access network with queue-aware transmission. [cs.IT], pp. 1–30 (2017). arXiv:1704.02902
14. Ephremides, A., Hajek, B.: Information theory and communication networks: an unconsummated union. IEEE Trans. Inf. Theor. **44**(6), 2416–2434 (1998)
15. Fanous, A., Ephremides, A.: Transmission control of two-user slotted ALOHA over Gilbert-Elliott channel: stability and delay analysis. In: Proceedings of the IEEE ISIT, St. Petersburg, Russia (2011)
16. Fayolle, G., Iasnogorodski, R., Malyshev, V.: Random walks in the quarter-plane, algebraic methods, boundary value problems and applications. Springer, Berlin (2017)
17. Fayolle, G., Iasnogorodski, R.: Two coupled processors: the reduction to a Riemann-Hilbert problem. Wahrscheinlichkeitstheorie **47**, 325–351 (1979)
18. Fiems, D., Phung-Duc, T.: Light-traffic analysis of queues with limited heterogenous retrials. In: QTNA2016, ACM, Wellington (2016)
19. Gakhov, F.D.: Boundary Value Problems. Pergamon Press, Oxford (1966)
20. Jeon, J., Ephremides, A.: On the stability of random multiple access with stochastic energy harvesting. IEEE J. Sel. Areas Commun. **33**(3), 571–584 (2015)
21. Jeon, J., Codreanu, M., Latva-aho, M., Ephremides, A.: The stability property of cognitive radio systems with imperfect sensing. IEEE J. Sel. Areas Commun. **32**(3), 628–640 (2014)
22. Kompella, S., Ephremides, A.: Stable throughput regions in wireless networks. Found. Trends Netw. **7**(4), 235–338 (2014)
23. Lau, C., Leung, C.: Capture models for mobile packet radio networks. IEEE Trans. Commun. **40**, 917–925 (1992)
24. Luo, W., Ephremides, A.: Stability of N interacting queues in random-access systems. IEEE Trans. Inf. Theor. **45**(5), 1579–1587 (1999)
25. Mahmoud, Q.: Cognitive Networks: Towards Self-Aware Networks. Wiley, Hoboken (2007)
26. Nain, P.: Analysis of a two-node Aloha network with infinite capacity buffers. In: Hasegawa, T., Takagi, H., Takahashi, Y. (eds.) Proceedings of the International Seminar on Computer Networking and Performance Evaluation, pp. 49–63, Tokyo (1985)
27. Naware, V., Mergen, G., Tong, L.: Stability and delay of finite-user slotted ALOHA with multipacket reception. IEEE Trans. Inf. Theor. **51**(7), 2636–2656 (2005)
28. Osseiran, A., et al.: Scenarios for 5G mobile and wireless communications: the vision of the METIS project. IEEE Commun. Mag. **52**(5), 26–35 (2014)
29. Pappas, N., Kountouris, M., Jeon, J., Ephremides, A., Traganitis, A.: Network-level cooperation in energy harvesting wireless networks. In: IEEE GlobalSIP, pp. 383–386, Austin, TX, USA (2013)
30. Pappas, N., Kountouris, M., Jeon, J., Ephremides, A., Traganitis, A.: Effect of energy harvesting on stable throughput in cooperative relay systems. J. Commun. Netw. **18**(2), 261–269 (2016)
31. Resing, J.A.C., Ormeci, L.: A tandem queueing model with coupled processors. Oper. Res. Lett. **31**, 383–389 (2003)
32. Rao, R., Ephremides, A.: On the stability of interacting queues in multiple access system. IEEE Trans. Inf. Theor. **34**, 918–930 (1989)
33. Sidi, M., Segall, A.: Two interfering queues in packet-radio networks. IEEE Trans. Commun. **31**(1), 123–129 (1983)
34. Szpankowski, W.: Stability Conditions for multidimensional queuing systems with applications. Oper. Res. **36**(6), 944–957 (1988)

35. Tsybakov, B.S., Mikhailov, V.A.: Ergodicity of a slotted ALOHA system. Probl. Peredachi Inf. **15**, 73–87 (1979)
36. Van Leeuwaarden, J.S.H., Resing, J.A.C.: A tandem queue with coupled processors: computational issues. Queueing Syst. **50**, 29–52 (2005)
37. Zorzi, M., Rao, R.: Capture and retransmission control in mobile radio. IEEE J. Sel. Areas Commun. **12**, 1289–1298 (1994)

Two-Way Communication M/M/1//N Retrial Queue

Velika Dragieva[1] and Tuan Phung-Duc[2(✉)]

[1] University of Forestry, 10 Kliment Ohridsky, 1756 Sofia, Bulgaria
dragievav@yahoo.com
[2] University of Tsukuba, 1-1-1 Tennodai, Tsukuba, Ibaraki 305-8573, Japan
tuan@sk.tsukuba.ac.jp

Abstract. We consider in this paper retrial queue with one server that serves a finite number of customers, each one producing a Poisson flow of incoming calls. In addition, after some exponentially distributed idle time the server makes outgoing calls of two types - to the customers in orbit and to the customers outside it. The outgoing calls of both types follow the same exponential distribution, different from the exponential service time distribution of the incoming calls. We derive formulas for computing the steady state distribution of the system state as well as formulas expressing the main performance macro characteristics in terms of the server utilization. Numerical examples are presented.

1 Introduction

Retrial queues of type $M/G/1//N$ in Kendall's notation are queueing models with 1 server which serves N customers (clients, calls) each one producing a Poisson flow of demands. Retrial feature is characterized by the specific behaviour of the arriving customers that find the server busy. These customers join a virtual waiting room, called orbit and repeat the attempt to get service after some time. The customers in the orbit are also called retrial customers or sources of retrial calls, while the customers that are not in the orbit or under service are called sources of primary calls or customers in free state.

Retrial queues arise from various real life situations as well as telecommunication and network systems (Falin and Templeton 1997; Artalejo and Gómez-Corral 2008). For example, in a call center a customer who cannot connect with an operator tries again later (Aguir et al. 2004). Furthermore, in modeling the mobile cellular systems, retrial feature cannot be ignored (Tran-Gia and Mandjes 1997; Van Do et al. 2014). The assumption of a finite number of customers is of special interest to practice, as in real situations the number of subscribers is finite. In particular, the described finite single server retrial queues and its variants are useful in modeling magnetic disk memory systems (Ohmura and Takahashi 1985), local area networks with nonpersistent CSMA/CD protocol (Li and Yang 1995), etc. Falin and Artalejo (1998) carried out an extensive analysis of the single server finite source retrial queue, including the busy period distribution and the waiting time process. Distribution of the number of retrials, made by a retrial customer while being in orbit is investigated by Dragieva (2013).

© Springer International Publishing AG 2017
N. Thomas and M. Forshaw (Eds.): ASMTA 2017, LNCS 10378, pp. 81–94, 2017.
DOI: 10.1007/978-3-319-61428-1_6

Single server finite source retrial queues with two types of breakdowns and repairs are considered by Wang et al. (2011) and by Zhang and Wang (2013).

There also exist real situations, especially in service systems where customers who cannot receive service immediately upon arrival register to the system and go to other places before returning to the system after some time. On the other hand, the server once becoming idle calls for customers. The former is reflected by retrials while the latter can be modelled by outgoing calls. These real situations are the motivation for us to consider finite source retrial models with two-way communication.

Some of the first results on two-way communication retrial queues are obtained by Falin (1979), who analyzes a single server queue in which the outgoing and the incoming calls are assumed to follow the same arbitrary service time distribution. The priority retrial queues with available buffers for the outgoing calls, studied by Falin et al. (1993) and by Choi et al. (1995) could also be considered as two-way communication models. Artalejo and Phung-Duc (2012) consider single and multiple servers retrial models with two-way communication where the service times of incoming and outgoing calls follow the exponential distribution with distinct parameters. The corresponding $M/G/1$ queue where the service times of incoming and outgoing calls follow two distinct arbitrary distributions is investigated by the same authors Artalejo and Phung-Duc (2013). Sakurai and Phung-Duc (2015) consider two-way communication retrial queues with multiple types of outgoing calls. A two-way communication $M/M/1$ retrial queue with server-orbit interaction is studied by Dragieva and Phung-Duc (2016). In the model, proposed in this paper it is assumed that after some exponentially distributed idle time the server makes outgoing calls of two types. The outgoing calls of type 1 are directed to the customers in orbit, while these that are of type 2 - to the customers outside the orbit. This assumption reflects various real-life situations, like call center of a credit card company where the operator may call to customers for some advertisement, or to the customers who not yet pay the money. But, at the moment some of these customers may be in the orbit. The operator is not notified for them, so that he/she may call to a customer outside the orbit as well as to a customer in orbit. Actually, when the population of customers is considered infinite, the probability that the server, calling to an arbitrary individual from this population may choose one from the orbit, is very small. Thus, in such situations it is more appropriate to model the system by queues with finite source. This motivated us to start investigation of finite source retrial models with two-way communication.

The rest of the current paper is organized as follows. In Sect. 2 we describe the model in detail. The joint distribution of the server state and the orbit size is studied in Sect. 3.1, while Sect. 3.2 deals with the main performance macro characteristics. Section 4 is devoted to numerical examples, Sect. 5 concludes the paper and presents some possible topics for future research.

2   Model Description

As stated in the Introduction we consider a queueing model with one server which serves N customers. Each of these customers produces a Poisson flow of incoming primary calls with mean $1/\lambda'$. Thus, when a source is free at time moment t (i.e. is not being served and is not waiting for service) it generates a primary call during an interval $(t, \ t+dt)$ with probability $\lambda'dt$. This means that if at a time moment t there are n customers in free state (sources of incoming primary calls), the arrival rate of the primary calls will be $n\lambda'$ and consequently the probability of a primary call arrival during a time interval $(t, \ t+dt)$ is equal to $n\lambda'dt$.

If an incoming call finds the server busy upon arrival it joins the orbit of retrial customers (calls), stays in it for an exponentially distributed time with mean $1/\mu$, and retries to get service. The incoming retrial (secondary) calls, like the incoming primary calls, are accepted if the server is idle, otherwise they enter the orbit again.

The server, in turn, after some exponentially distributed idle time makes outgoing calls of two types - to a customer in the orbit (an outgoing call of type 1) or to a customer in free state (an outgoing call of type 2). The parameters of these exponential distributions are α and α_0', respectively. Thus, if the server is idle and there are n incoming retrial customers in the orbit, the server connects with one of them in an exponentially distributed time with parameter $n\alpha$, and connects with one of the customers outside the orbit in an exponentially distributed time with parameter $(N-n)\alpha_0'$.

The service times of the incoming calls and the outgoing calls of both types are exponentially distributed with rate ν_1 and ν_2, respectively. When the service is over all customers go to a free state.

We assume that the arrivals of primary incoming calls, retrial intervals of secondary incoming calls, service times of incoming and outgoing calls, and the time to make outgoing calls are mutually independent.

We denote the number of customers in orbit, the server state and the number of busy servers at time t by $R(t)$, $S(t)$ and $C(t)$, respectively,

$$S(t) = \begin{cases} 0, & \text{when the server is idle,} \\ 1, & \text{when an incoming call is in service,} \\ 2, & \text{when an outgoing call is in service,} \end{cases}$$

$$C(t) = \begin{cases} 0, & \text{when } S(t) = 0, \\ 1, & \text{when } S(t) = 1, 2. \end{cases}$$

Obviously, when the server is busy the number of customers in the orbit can't be equal to N, i.e. $R(t) < N$. As stated above after the service both incoming and outgoing customers go to a free state. This means that when the server is idle, there will be at least one customer in free state, i.e. again $R(t) < N$. Thus, the state space of the process $(S(t), R(t))$ is the set $\{0, 1, 2\} \times \{0, 1, 2, \dots, N-1\}$. Because of the finite state space this Markovian process is always stable.

Some particular values of the parameters in the above described model lead to other models. Namely, in the case

- $\alpha = \alpha_0' = 0$ we obtain the classical single server retrial queue, studied by a number of authors, in a number of papers, some of which are presented in the Introduction;

- $\alpha_0' = 0$ we have a single server, finite source retrial queue with search of the customers from orbit;
- $\mu = \lambda'$ we get a single server, finite source queue with losses and two-way communication.

Finally, if $N \to \infty$, $\lambda' \to 0$ and $\alpha_0' \to 0$ in such a way that $N\lambda' \to \lambda$ and $N\alpha_0' \to \alpha_0$ our model converges to the corresponding model with infinite source, studied by Dragieva and Phung-Duc (2016).

Further in the paper we discuss some of these particular cases.

3   Stationary System State Distributions

3.1   Joint Distribution of the Server State and Orbit Size

The system of balance equations for the stationary probabilities

$$\pi_{i,j} = \lim_{t \to \infty} P\left(S(t) = i, R(t) = j\right) \quad i = 0, 1, 2, \; j = 0, 1, \ldots, N-1$$

is

$$[(N-j)\left(\lambda' + \alpha_0'\right) + j(\alpha + \mu)]\,\pi_{0,j} = \nu_1 \pi_{1,j} + \nu_2 \pi_{2,j}, \tag{1}$$

$$[(N-j-1)\lambda' + \nu_1]\,\pi_{1,j} = (N-j)\lambda' \pi_{0,j} + (j+1)\mu \pi_{0,j+1} + (N-j)\lambda' \pi_{1,j-1}, \tag{2}$$

$$[(N-j-1)\lambda' + \nu_2]\,\pi_{2,j} = (N-j)\alpha_0' \pi_{0,j} + (j+1)\alpha \pi_{0,j+1} + (N-j)\lambda' \pi_{2,j-1}, \tag{3}$$

with $\pi_{0,N} = \pi_{1,-1} = \pi_{2,-1} = 0$.

Because of the finite number of equations we can solve this system using general methods like Cramer's rule. But here we present more convenient recursive schemes. Firstly, they can save a number of operations, and secondly will be useful in our future work when investigating the other descriptors of the system functioning, like the waiting time process, busy period distribution and others. We first express the probabilities $\pi_{i,j}$ ($i = 1, 2$) in terms of the probabilities $\pi_{0,j}(j = 0, 1, \ldots, N-1)$. According to Eqs. (2) and (3), if denote

$$\pi_{i,j} = A_{j,0}^{(i)} \pi_{0,0} + A_{j,1}^{(i)} \pi_{0,1} + \ldots + A_{j,j+1}^{(i)} \pi_{0,j+1}, \tag{4}$$

we have

$$\pi_{1,0} = \frac{N\lambda' \pi_{0,0}}{a_{1,1}} + \frac{\mu \pi_{0,1}}{a_{1,1}}, \quad \pi_{2,0} = \frac{N\alpha_0' \pi_{0,0}}{a_{2,1}} + \frac{\alpha \pi_{0,1}}{a_{2,1}},$$

$$\pi_{1,j} = \frac{(N-j)\lambda'}{a_{1,j+1}} \sum_{k=0}^{j} A_{j-1,k}^{(1)} \pi_{0,k} + \frac{(N-j)\lambda'}{a_{1,j+1}} \pi_{0,j} + \frac{(j+1)\mu}{a_{1,j+1}} \pi_{0,j+1},$$

$$\pi_{2,j} = \frac{(N-j)\lambda'}{a_{2,j+1}} \sum_{k=0}^{j} A_{j-1,k}^{(2)} \pi_{0,k} + \frac{(N-j)\alpha_0'}{a_{2,j+1}} \pi_{0,j} + \frac{(j+1)\alpha}{a_{2,j+1}} \pi_{0,j+1},$$

$j = 1, \ldots, N-1$, where

$$a_{i,j} = (N-j)\lambda' + \nu_i, \ \pi_{i,N} = 0, \ j = 1, \ldots, N, \ i = 1, 2.$$

This gives the following recursive formulas for calculation of the coefficients $A_{j,k}^{(i)}$:

$$A_{j,k}^{(1)} = \frac{(N-j)\lambda'}{a_{1,j+1}} \left(A_{j-1,k}^{(1)} + \delta_{k,j} \right), \tag{5}$$

$$A_{j,k}^{(2)} = \frac{(N-j)}{a_{2,j+1}} \left(\lambda' A_{j-1,k}^{(2)} + \delta_{k,j} \alpha_0' \right), \ k = 0, \ldots, j, \tag{6}$$

$$A_{j,j+1}^{(1)} = \frac{(j+1)\mu}{a_{1,j+1}}, \ A_{j,j+1}^{(2)} = \frac{(j+1)\alpha}{a_{2,j+1}}, \ j = 0, 1, \ldots, N-1, \tag{7}$$

$$A_{-1,0}^{(1)} = A_{-1,0}^{(2)} = A_{N-1,N}^{(1)} = A_{N-1,N}^{(2)} = 0.$$

Here $\delta_{k,j}$ is the Kronecker's symbol, which is equal to 1 if $k = j$, and is equal to 0 if $k \neq j$.

The explicit expressions, based on these recursive formulas are:

$$A_{j,k}^{(1)} = \frac{(N-k)\ldots(N-j)(\lambda')^{j+1-k}(a_{1,k}+k\mu)}{a_{1,k}a_{1,k+1}a_{1,k+2}\cdots a_{1,j+1}}, \tag{8}$$

$$A_{j,k}^{(2)} = \frac{(N-k)\ldots(N-j)(\lambda')^{j-k}(a_{2,k}\alpha_0'+k\lambda'\alpha)}{a_{2,k}a_{2,k+1}a_{2,k+2}\cdots a_{2,j+1}}, \tag{9}$$

for $j = 0, 1, \ldots, N-1$, $k = 0, \ldots, j$, $a_{1,0} = a_{2,0} = 1$. The expressions for $A_{j,j+1}^{(i)}$ $(i = 1, 2)$ are given by (7).

Next, in Eq. (1) we substitute $\pi_{i,j}$ according to formulas (4), $(j=0, \ldots, N-2)$ and obtain a relation between the probabilities $\pi_{0,j}$:

$$\left[\nu_1 A_{j,j+1}^{(1)} + A_{j,j+1}^{(2)} \nu_2 \right] \pi_{0,j+1} =$$
$$\left[(N-j)(\lambda'+\alpha_0') + j(\alpha+\mu) - \left(\nu_1 A_{j,j}^{(1)} + A_{j,j}^{(2)} \nu_2 \right) \right] \pi_{0,j} -$$
$$\left[\left(\nu_1 A_{j,0}^{(1)} + A_{j,0}^{(2)} \nu_2 \right) \pi_{0,0} + \ldots + \left(\nu_1 A_{j,j-1}^{(1)} + A_{j,j-1}^{(2)} \nu_2 \right) \pi_{0,j-1} \right], \tag{10}$$
$$j = 0, 1, \ldots, N-2, \ A_{0,-1}^{(1)} = A_{0,-1}^{(2)} = 0.$$

Following this scheme we can express all probabilities $\pi_{0,j}$ $(j = 1, \ldots, N-1)$ in terms of $\pi_{0,0}$. Then, from (4) we can express $\pi_{i,j}$ $(i = 1, 2, \ j = 0, 1, \ldots, N-1)$ also in terms of $\pi_{0,0}$. Finally, from the normalizing condition

$$\sum_{i=0}^{2} \sum_{j=0}^{N-1} \pi_{ij} = 1$$

we can find $\pi_{0,0}$. Thus, we can calculate the stationary system state distribution.

Further, having $\pi_{0,0}$ found we can calculate the distribution $\pi_{i,j}$ not only using formulas (10) and (4), but also by the recursive formulas, presented in the next Proposition.

Proposition 1. *The stationary joint distribution $\pi_{i,j}$ of the server state and the orbit size satisfies the following recursive formulas:*

$$\pi_{0,j} = \frac{(N-j)\lambda'}{j(\mu+\alpha)}(\pi_{1,j-1}+\pi_{2,j-1}), \quad j=1,\ldots,N-1, \tag{11}$$

$$\begin{aligned}\pi_{1,j} = \frac{(N-j)\lambda'}{(N-j-1)\lambda'(\mu\nu_1+\alpha\nu_2)+\nu_1\nu_2(\alpha+\mu)}\{(N-j-1)\lambda'\mu\pi_{2,j-1}+\\ [\nu_2(\alpha+\mu)+(N-j-1)\mu(\lambda'+\alpha_0')]\pi_{0,j}+\\ [(N-j-1)\lambda'\mu+\nu_2(\alpha+\mu)]\pi_{1,j-1}\},\end{aligned} \tag{12}$$

$$\begin{aligned}\pi_{2,j} = \frac{(N-j)}{(N-j-1)\lambda'(\mu\nu_1+\alpha\nu_2)+\nu_1\nu_2(\alpha+\mu)}\{(N-j-1)(\lambda')^2\alpha\pi_{1,j-1}+\\ [(N-j-1)\lambda'\alpha(\lambda'+\alpha_0')+\alpha_0'\nu_1(\alpha+\mu)]\pi_{0,j}+\\ \lambda'[(N-j-1)\lambda'\alpha+\nu_1(\alpha+\mu)]\pi_{2,j-1}\}.\end{aligned} \tag{13}$$

Proof. We sum up Eqs. (1)–(3) for $j=0$ and obtain formula (11) for $j=1$,

$$(N-1)\lambda'(\pi_{1,0}+\pi_{2,0}) = (\mu+\alpha)\pi_{0,1}. \tag{14}$$

Then we sum Eqs. (1)–(3) for $j=1$,

$$(\mu+\alpha)\pi_{0,1}+(N-2)\lambda'(\pi_{1,0}+\pi_{2,0}) =$$
$$2(\mu+\alpha)\pi_{0,2}+(N-1)\lambda'(\pi_{1,0}+\pi_{2,0}),$$

and add it to (14). Thus we get (11) for $j=2$. Further, by induction it is easy to prove relations (11) for all $j=1,\ldots,N-1$. The rest of the recurrent formulas (11)–(13) follow from the combination of (2)–(3) with (11). Namely, substituting $\pi_{0,j+1}$ from (11) into (2) and (3), after some transformations we get

$$\frac{(N-j-1)\lambda'\alpha+\nu_1(\alpha+\mu)}{\mu+\alpha}\pi_{1,j} =$$
$$(N-j)\lambda'\pi_{0,j}+\frac{(N-j-1)\lambda'\mu}{\mu+\alpha}\pi_{2,j}+(N-j)\lambda'\pi_{1,j-1},$$
$$\frac{(N-j-1)\lambda'\mu+\nu_2(\alpha+\mu)}{\mu+\alpha}\pi_{2,j} =$$
$$(N-j)\alpha_0'\pi_{0,j}+\frac{(N-j-1)\lambda'\alpha}{\mu+\alpha}\pi_{1,j}+(N-j)\lambda'\pi_{2,j-1}.$$

Now we substitute $\pi_{2,j}$ from the second into the first of these equations, and $\pi_{1,j}$ from the first into the second,

$$\frac{(N-j-1)\lambda'\alpha+\nu_1(\alpha+\mu)}{\mu+\alpha}\pi_{1,j} =$$
$$(N-j)\lambda'\pi_{0,j}+(N-j)\lambda'\pi_{1,j-1}+\frac{(N-j-1)\lambda'\mu}{[(N-j-1)\lambda'\mu+\nu_2(\alpha+\mu)]}\times$$
$$\left[(N-j)\alpha_0'\pi_{0,j}+\frac{(N-j-1)\lambda'\alpha}{\mu+\alpha}\pi_{1,j}+(N-j)\lambda'\pi_{2,j-1}\right],$$
$$\frac{(N-j-1)\lambda'\mu+\nu_2(\alpha+\mu)}{\mu+\alpha}\pi_{2,j} =$$
$$(N-j)\alpha_0'\pi_{0,j}+(N-j)\lambda'\pi_{2,j-1}+\frac{(N-j-1)\lambda'\alpha}{(N-j-1)\lambda'\alpha+\nu_1(\alpha+\mu)}\times$$
$$\left[(N-j)\lambda'\pi_{0,j}+\frac{(N-j-1)\lambda'\mu}{\mu+\alpha}\pi_{2,j}+(N-j)\lambda'\pi_{1,j-1}\right].$$

Rearranging the terms, the last two equations give formulas (12) and (13).

Remark 1. If in formulas (11)–(13) we fix j and take limits as $N \to \infty$, $\lambda' \to 0$ and $\alpha'_0 \to 0$ in such a way that $N\lambda' \to \lambda$, $N\alpha'_0 \to \alpha_0$, we obtain exactly the recurrent formulas, connecting the stationary system state probabilities for the corresponding model with infinite source (Proposition 2 in Dragieva and Phung-Duc (2016)). Similarly, if take $\alpha = \alpha'_0 = 0$, then formulas (11)–(12) give the recursive formulas, obtained by Dragieva (2013) for the corresponding finite source retrial queue without two-way communication.

Remark 2. In fact, we do not need recursions (11)–(13) for the calculation of the system state distribution because we have the more convenient formulas (10), (5)–(7) and (4). Exactly these formulas are used in the calculation of numerical examples, presented in Sect. 4. Recursive formulas (11)–(13) may be useful in the analysis of the waiting time process, analogously to the corresponding formulas in the single server, finite source retrial queue without two-way communication (see Dragieva 2013).

Now we turn attention to the system state distribution at the moments of a primary incoming call arrival. In the models with finite source this distribution differs from the corresponding distribution at any arbitrary time moment (which is discussed in detail for example in Falin and Artalejo (1998) or in Dragieva (2013)). Thus, if we introduce the event $A(t)$ that at time t a primary call arrives and denote by $\overline{\pi}_{i,j}$ the stationary conditional probabilities

$$\overline{\pi}_{i,j} = \lim_{t \to \infty} P\left\{ S(t) = i, R(t) = j \,|\, A(t) \right\} \quad i = 0, 1, 2, \ j = 0, 1, \ldots, N-1,$$

then

$$\overline{\pi}_{i,j} = \begin{cases} \frac{(N-j)\lambda' \pi_{0,j}}{D}, & \text{if } i = 0, \\ \frac{(N-j-1)\lambda' \pi_{i,j}}{D}, & \text{if } i = 1, 2, \end{cases} \tag{15}$$

$$D = \sum_{i=1}^{2} \sum_{n=0}^{N-1} (N-n-1)\lambda' \pi_{i,n} + \sum_{n=0}^{N-1} (N-n)\lambda' \pi_{0,n}. \tag{16}$$

This distribution is important in the investigation of the waiting time process.

3.2   Main Macro Characteristics of the System Performance

In the models with finite state space, if the system state distribution is obtained, then it is not difficult to calculate any of the basic macro characteristics of the system performance. Nevertheless, here we derive formulas, expressing these characteristics in terms of the server utilization (or the idle server probability).

Let's denote

$$P_i = \lim_{t \to \infty} P\left\{ S(t) = i \right\} = \sum_{j=0}^{N-1} \pi_{i,j},$$

$$M_{i,p} = \sum_{j=0}^{N-1} j^p \pi_{i,j}, p = 0, 1, \ldots, \ i = 0, 1, 2,$$

$$P_i = M_{i,0}.$$

Summing all Eq. (2), then (3) over $j = 0, \ldots, N - 1$, we obtain equations for the stationary server state distribution P_i $(i = 0, 1, 2)$ and the first partial moment $M_{0,1}$,

$$\nu_1 P_1 = N\lambda' P_0 + (\mu - \lambda') M_{0,1}, \tag{17}$$

$$\nu_2 P_2 = N\alpha_0' P_0 + (\alpha - \alpha_0') M_{0,1}. \tag{18}$$

As stated in Sect. 2, if $\mu = \lambda'$ we have no orbit and the model is modified to the particular case of a finite source queue with losses and two-way communication. In this case it is reasonable to take $\alpha = \alpha_0'$, but there are real situations when we can consider $\alpha \neq \alpha_0'$. For example, in a call center of some company the operator can record all unsuccessful clients and although they give up their request $(\mu = \lambda')$ he/she can call to them for advertising, reminders, or anything else, with specific intensity $(\alpha \neq \alpha_0')$. In the case $\mu = \lambda'$ and $\alpha = \alpha_0'$, from (17), (18) and the normalizing condition

$$P_0 + P_1 + P_2 = 1$$

we obtain formulas for the probabilities P_i $(i = 0, 1, 2)$:

$$P_1 = \frac{N\lambda'}{\nu_1} P_0, \; P_2 = \frac{N\alpha}{\nu_2} P_0, \; P_0 = \frac{\nu_1 \nu_2}{N(\lambda' \nu_2 + \alpha \nu_1) + \nu_1 \nu_2}.$$

Further we assume that either $\mu \neq \lambda'$ or $\alpha \neq \alpha_0'$. Equations (17), (18) and the normalizing condition allow to express P_i $(i = 1, 2)$ and $M_{0,1}$ in terms of P_0:

$$P_1 = \frac{(\mu - \lambda') \nu_2 (1 - P_0) + N(\alpha\lambda' - \alpha_0'\mu) P_0}{(\alpha - \alpha_0') \nu_1 + (\mu - \lambda') \nu_2}, \tag{19}$$

$$P_2 = \frac{(\alpha - \alpha_0') \nu_1 (1 - P_0) - N[\alpha\lambda' - \alpha_0'\mu] P_0}{(\alpha - \alpha_0') \nu_1 + (\mu - \lambda') \nu_2}, \tag{20}$$

$$M_{0,1} = \frac{\nu_1 \nu_2 (1 - P_0) - N(\alpha_0'\nu_1 + \lambda'\nu_2) P_0}{(\alpha - \alpha_0') \nu_1 + (\mu - \lambda') \nu_2}. \tag{21}$$

Further, multiplying Eq. (1) by j $(j = 1, \ldots, N - 1)$, then (2) and (3) by $(j + 1)$ $(j = 0, 1, \ldots, N - 1)$ and summing each of these three groups equations over j we get relations between $M_{i,0} = P_i$, $M_{i,1}$, $(i = 0, 1, 2)$, and $M_{0,2}$,

$$N(\lambda' + \alpha_0') M_{0,1} + (\alpha + \mu - \lambda' - \alpha_0') M_{0,2} = \nu_1 M_{1,1} + \nu_2 M_{2,1}, \tag{22}$$

$$(\nu_1 + \lambda') M_{1,1} + [\nu_1 - \lambda'(N - 1)] P_1 = \\ N\lambda' P_0 + (N - 1)\lambda' M_{0,1} + (\mu - \lambda') M_{0,2}, \tag{23}$$

$$(\nu_2 + \lambda') M_{2,1} + [\nu_2 - \lambda'(N - 1)] P_2 = \\ N\alpha_0' P_0 + (N - 1)\alpha_0' M_{0,1} + (\alpha - \alpha_0') M_{0,2}. \tag{24}$$

These equations allow to express the partial moments $M_{0,2}, M_{1,1}$ and $M_{2,1}$ in terms of $M_{0,0} = P_0$. Thus, the mean orbit size also can be expressed in terms of P_0.

Proposition 2. *The mean orbit size, M_1 is equal to*

$$M_1 = \lim_{t \to \infty} E\left[R(t)\right] = \sum_{i=0}^{2} \sum_{j=1}^{N-1} j\pi_{i,j} =$$

$$M_{0,1} + M_{1,1} + M_{2,1} = N - 1 + P_0 + \tag{25}$$

$$\frac{N\left[(\alpha+\mu)\alpha_0'\nu_1 + \alpha\lambda'(\nu_2-\nu_1)\right]P_0 - \left(\mu-\lambda'+\alpha\right)\nu_1\nu_2(1-P_0)}{\lambda'\left[(\alpha-\alpha_0')\nu_1 + (\mu-\lambda')\nu_2\right]}.$$

Proof. From Eq. (22) we express $M_{0,2}$ in terms of $M_{i,1}$ $(i = 0, 1, 2)$

$$M_{0,2} = \frac{\nu_1 M_{1,1} + \nu_2 M_{2,1} - N\left(\lambda' + \alpha_0'\right)M_{0,1}}{\alpha + \mu - \lambda' - \alpha_0'}, \quad \alpha + \mu \neq \lambda' + \alpha_0',$$

and substitute it in the Eqs. (23) and (24). After some transformations this leads to the following system for $M_{i,1}$ $(i = 1, 2)$

$$\left(\lambda' + \frac{\nu_1(\alpha-\alpha_0')}{\alpha+\mu-\lambda'-\alpha_0'}\right)M_{1,1} - \frac{(\mu-\lambda')\nu_2}{\alpha+\mu-\lambda'-\alpha_0'}M_{2,1} =$$

$$N\lambda'P_0 - \left[\nu_1 - \lambda'(N-1)\right]P_1 + \left(-\lambda' + \frac{N(\lambda'\alpha-\alpha_0'\mu)}{\alpha+\mu-\lambda'-\alpha_0'}\right)M_{0,1},$$

$$-\frac{(\alpha-\alpha_0')\nu_1}{\alpha+\mu-\lambda'-\alpha_0'}M_{1,1} + \left(\lambda' + \frac{\nu_2(\mu-\lambda')}{\alpha+\mu-\lambda'-\alpha_0'}\right)M_{2,1} =$$

$$N\alpha_0'P_0 - \left[\nu_2 - \lambda'(N-1)\right]P_2 + \left[-\alpha_0' + \frac{N(\alpha_0'\mu-\lambda'\alpha)}{\alpha+\mu-\lambda'-\alpha_0'}\right]M_{0,1}.$$

Summing up these two equations we get

$$\lambda'\left(M_{1,1} + M_{2,1}\right) = N\left(\lambda' + \alpha_0'\right)P_0 -$$
$$(\lambda' + \alpha_0')M_{0,1} - \left\{\left[\nu_1 - \lambda'(N-1)\right]P_1 + \left[\nu_2 - \lambda'(N-1)\right]P_2\right\}.$$

Thus, for the mean orbit size it holds

$$M_1 = M_{0,1} + M_{1,1} + M_{2,1} =$$
$$\frac{N(\lambda'+\alpha_0')P_0}{\lambda'} - \frac{\alpha_0'}{\lambda'}M_{0,1} -$$
$$\frac{\left[\nu_1-\lambda'(N-1)\right]P_1 + \left[\nu_2-\lambda'(N-1)\right]P_2}{\lambda'}.$$

Substituting here P_i $(i = 1, 2)$ and $M_{0,1}$ with the expressions (19)–(21) we obtain formula (25).

Using formulas (19)–(21) and (25) we can express the other basic performance measures:

- The blocking probability P_B that an arriving primary incoming call will be blocked in the orbit of retrial customers,

$$P_B = \sum_{i=1}^{2} \sum_{n=0}^{N-1} \overline{\pi}_{i,n} =$$

$$\frac{\sum_{i=1}^{2}\sum_{n=0}^{N-1}(N-n-1)\lambda'\pi_{i,n}}{\sum_{i=1}^{2}\sum_{n=0}^{N-1}(N-n-1)\lambda'\pi_{i,n}+\sum_{n=0}^{N-1}(N-n)\lambda'\pi_{0,n}} =$$

$$\frac{(N-1)\lambda'(1-P_0)-\lambda'(M_{1,1}+M_{2,1})}{N\lambda'-\lambda'(1-P_0+M_1)} = 1 + \frac{M_{0,1}-NP_0}{N-(1-P_0+M_1)};$$

- Mean rate of generation of primary incoming calls,

$$\Lambda = \lambda' \lim_{t \to \infty} E\left[N - C(t) - R(t)\right] = \lambda'\left[N - (P_1 + P_2) - M_1\right] =$$

$$\frac{(\mu - \lambda' + \alpha)\,\nu_1\nu_2\,(1 - P_0) - N\left[(\alpha + \mu)\,\alpha_0'\nu_1 + \alpha\lambda'\,(\nu_2 - \nu_1)\right] P_0}{(\alpha - \alpha_0')\,\nu_1 + (\mu - \lambda')\,\nu_2};$$

- Mean value of the waiting time $W(t)$, that a primary incoming call, arriving at time moment t will spend in the orbit. Using Little's formula for this mean value we have:

$$\lim_{t \to \infty} E[W(t)] = \Lambda^{-1} \lim_{t \to \infty} E\left[R(t)\right] = \frac{M_1}{\Lambda} =$$

$$\frac{(N-1+P_0)\left[(\alpha-\alpha_0')\nu_1+(\mu-\lambda')\nu_2\right]}{N\left[(\alpha+\mu)\alpha_0'\nu_1+\alpha\lambda'(\nu_2-\nu_1)\right]P_0-(\mu-\lambda'+\alpha)\nu_1\nu_2(1-P_0)} - \frac{1}{\lambda'};$$

- Mean number $E\left[RA(t)\right]$ of retrial attempts, that a primary incoming customer, arriving at time moment t will make while being in the orbit. If the intensity of the outgoing calls to the customers in orbit is 0 ($\alpha = 0$) then the following relation holds:

$$\lim_{t \to \infty} E\left[RA(t)\right] = \mu \lim_{t \to \infty} E\left[W(t)\right] = \mu \frac{M_1}{\Lambda}.$$

In the case $\alpha > 0$ if we want to calculate the mean number of retrials, we should investigate the stationary distribution of this number, which will be one of our future work.

Remark 3. For $\alpha = \alpha_0' = 0$ all formulas, obtained in this Section coincide with the formulas derived by Falin and Artalejo (1998) for the corresponding model without two-way communication.

4   Numerical Examples

In this Section we present numerical examples, illustrating the influence of the system parameters on the main performance macro characteristics, considered in previous Section.

Figure 1 shows the dependence of the stationary server state distribution P_i ($i = 0, 1, 2$) on the parameters λ' (left upper corner), α_0' (right upper corner), ν_1 (left lower corner) and N (right lower corner). We see that the behaviour of most of the presented functions is intuitively expected:

- The proportion of time P_1 that the server is busy with incoming calls service increases with the increase of primary intensity λ' and the mean service time of incoming calls, $1/\nu_1$. P_1 decreases with increase of the intensity α_0' of the outgoing calls to the customers in free state. Numerical examples, not presented here show that P_1 increases with the increase of the secondary intensity, μ and decreases with the increase of the mean service time of outgoing calls, $1/\nu_2$ and with the increase of the intensity of outgoing calls to the orbit, α.

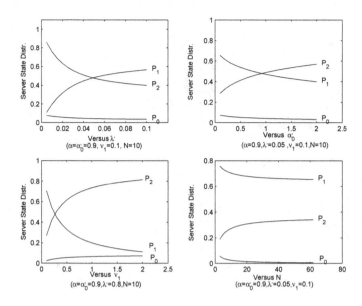

Fig. 1. Stationary server state distribution $P_i = P(S = i)$ $(i = 0, 1, 2)$ versus system parameters λ', α_0', ν_1, N. ($\mu = 0.2$, $\nu_2 = 0.8$)

- All presented examples show that when P_1 increases, then the proportion of time P_2 that the server is busy with outgoing calls decreases and vice versa.

It is interesting that for all presented values of the system parameters the server utilization, $P_1 + P_2 = 1 - P_0$ is almost equal to 1. The increase of the number N of all clients of the system has little impact on the server state distribution, keeping P_1 greater than P_2 and P_0, the last one almost equal to 0.

The dependence of the rest of the macro characteristics (the first partial moments, $M_{i,1}$, $(i = 0, 1, 2)$ and mean orbit size, M_1, blocking probability, P_B, mean waiting time, $E[W] = \lim_{t \to \infty} E[W(t)]$ and mean rate of generation of primary incoming calls, Λ) on the system parameters follow intuitively expected behaviour. The only exception is the primary incoming calls intensity λ', which influence on the system performance is hard to be intuitively explained. To show this influence in more detail we present it in two figures - Figs. 2 and 3. On these figures we can see that for all presented values of the system parameters the blocking probability confirms the well known property of the finite source retrial queues to have a point of maximum as a function of λ' (Falin and Artalejo 1998; Almaási et al. 2005; Wang et al. 2011; Zhang and Wang 2013). The new comes with the behaviour of the mean waiting time, $E[W]$ and the mean rate of generation of primary incoming calls, Λ. We see in Fig. 2 that there exist values of the system parameters for which, like in the other finite source retrial queues (Falin and Artalejo 1998; Almaási et al. 2005; Wang et al. 2011; Zhang and Wang 2013), $E[W]$ has a point of maximum. But, on Fig. 3 we see that for the same values of these parameters it has and a point of minimum. This property has

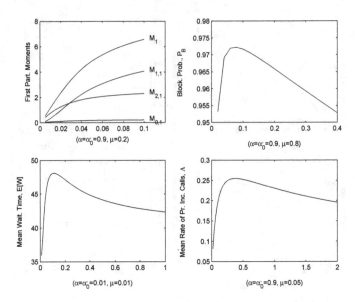

Fig. 2. Basic performance macro characteristics versus primary intensity λ'. ($\nu_1 = 0.1, \nu_2 = 0.8, N = 10$)

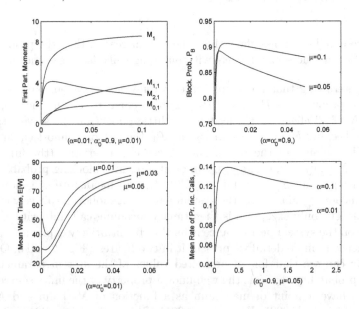

Fig. 3. Basic performance macro characteristics versus primary intensity λ'. ($\nu_1 = 0.1, \nu_2 = 0.8, N = 10$)

not been observed till now in the related literature. We also see on Fig. 3 that it does not hold for all values of the system parameters. Analogously, we see that the behaviour of Λ as a function of λ' also depends on the values of the rest of the system parameters - for some of them it is a strictly increasing function, but for some of them it has a point of maximum. The last property has not been observed till now. For example, all numerical results, presented by Wang et al. (2011) for the single server, finite source retrial queue with server breakdowns and repairs show that Λ follows the intuitively expected behaviour to be strictly increasing as a function of the primary intensity λ'.

It is interesting to note that for the values of the system parameters, presented on Fig. 2, the partial moment $M_{2,1}$ is an increasing function of λ', but for the values, presented on Fig. 3 it has a point of maximum.

5   Conclusion and Future Work

In this paper we derive formulas for the joint distribution of the server state and the orbit size in a finite source retrial queue of $M/M/1//N$ type with two-way communication. Main performance macro characteristics are expressed in terms of the server utilization. The influence of the system parameters on these macro characteristics is studied on the basis of numerical examples. Formulas, obtained in the present paper allow to extend this investigation by studying the waiting time process, the busy period distribution and other descriptors of the system performance like the number of successful and blocked events. We also plan to consider the corresponding queue of type $M/G/1//N$.

Acknowledgements. The authors would like to thank anonymous referees for their constructive comments which improved the presentation of the paper.

References

Aguir, S., Karaesmen, E., Aksin, O., Chauvet, F.: The impact of retrials on call center performance. OR Spectr. **26**, 353–376 (2004)

Almaási, B., Roszik, J., Sztrik, J.: Homogeneous finite-source retrial queues with server subject to breakdowns and repairs. Math. Comput. Model. **42**, 673–682 (2005)

Artalejo, J., Gómez-Corral, A.: Retrial Queueing Systems: A Computational Approach. Springer, Heidelberg (2008)

Artalejo, J., Phung-Duc, T.: Markovian retrial queues with two way communication. J. Ind. Manag. Optim. **8**, 781–806 (2012)

Artalejo, J., Phung-Duc, T.: Single server retrial queues with two way communication. Appl. Math. Model. **37**, 1811–1822 (2013)

Choi, B., Choi, K., Lee, Y.: M/G/1 retrial queueing systems with two types of calls and finite capacity. Queueing Syst. **19**, 215–229 (1995)

Dragieva, V.: A finite source retrial queue: number of retrials. Commun. Stat. - Theory Methods **42**(5), 812–829 (2013)

Dragieva, V., Phung-Duc, T.: Two-way communication M/M/1 retrial queue with server-orbit interaction. In: Proceedings of the 11th International Conference on Queueing Theory and Network Applications, (ACM Digital Library), 7 p. (2016). doi:10.1145/3016032.3016049

Falin, G.: Model of coupled switching in presence of recurrent calls. Eng. Cybern. Rev. **17**, 53–59 (1979)

Falin, G., Artalejo, J., Martin, M.: On the single server retrial queue with priority customers. Queueing Syst. **14**, 439–455 (1993)

Falin, G., Templeton, J.: Retrial Queues. Chapman and Hall, London (1997)

Falin, G., Artalejo, J.: A finite source retrial queue. Eur. J. Oper. Res. **108**, 409–424 (1998)

Li, H., Yang, T.: A single server retrial queue with server vacations and a finite number of input sources. Eur. Oper. Res. **85**, 149–160 (1995)

Ohmura, H., Takahashi, Y.: An analysis of repeated call model with a finite number of sources. Electron. Commun. Jpn. **68**, 112–121 (1985)

Sakurai, H., Phung-Duc, T.: Two-way communication retrial queues with multiple types of outgoing calls. Top **23**, 466–492 (2015)

Tran-Gia, P., Mandjes, M.: Modeling of customer retrial phenomenon in cellular mobile networks. IEEE J. Sel. Areas Commun. **15**, 1406–1414 (1997)

Zhang, F., Wang, J.: Performance analysis of the retrial queues with finite number of sources and server interruptions. J. Korean Stat. Soc. **42**, 117–131 (2013)

Van Do, T., Wochner, P., Berches, T., Sztrik, J.: A new finite-source queueing model for mobile cellular networks applying spectrum renting. Asia - Pac. J. Oper. Res. **31**, 14400004 (2014)

Wang, J., Zhao, L., Zhang, F.: Analysis of the finite source retrial queues with server breakdowns and repairs. J. Ind. Manag. Optim. **7**(3), 655–676 (2011)

An Algorithmic Approach for Multiserver Retrial Queues with Two Customers Classes and Non-preemptive Priority

Nawel Gharbi[1]([⊠]) and Leila Charabi[2]

[1] Department of Computer Science,
University of Sciences and Technology USTHB, Algiers, Algeria
ngharbi@usthb.dz
[2] National Computer Science Engineering School ESI, Algiers, Algeria
l_charabi@esi.dz

Abstract. Retrial queueing models with multiple servers and two classes of customers arise in various practical computer and telecommunication systems. The consideration of retrials (or repeated attempts) introduces analytical difficulties and most of works consider either models with preemptive priority or non-preemptive priority in the single server case. This paper aims to propose a recursive algorithmic approach for the performance analysis of a multiserver retrial queue with non-preemptive priority and two customers classes: ordinary customers whose access to the service depends on the number of available servers and who join the orbit when blocked; and impatient priority customers who have access to all servers and are lost when no server is available. In addition, we develop the formula of the main stationary performance measures. Through numerical examples, we study the effect of the system parameters on the blocking probability for ordinary customers and the loss probability for priority customers.

Keywords: Retrial multiserver queues · Two customers classes · Impatient customers · Non-preemptive priority · Recursive algorithm · Performance measures

1 Introduction

Retrial queueing models are characterized by the feature of retrial phenomenon that an arriving customer who finds all servers (or resources) occupied, joins the virtual group of blocked customers, called *orbit* and retry again for service after a random amount of time. Models with retrials have been widely used to analyze several practical problems in computer networks, telecommunication systems, call centers, cellular mobile networks [1–5] and wireless sensor networks [6]. Significant references and important surveys on this topic [7–9] reveal the non-negligible impact of retrials, which arise due to a blocking in a system with limited capacity resources or due to impatience of customers.

© Springer International Publishing AG 2017
N. Thomas and M. Forshaw (Eds.): ASMTA 2017, LNCS 10378, pp. 95–108, 2017.
DOI: 10.1007/978-3-319-61428-1_7

In fact, the consideration in the modeling process of the customers behavior, especially, retrials of customers whose request was rejected because of the lack of available resources, is crucial to determine the system performances, because it's well known that the retrials can negatively affect the system performance, because they generate more load. However, the consideration of retrial phenomenon introduces great analytical complications to obtain most important performance indices. In particular, for multiserver models, no explicit closed-form solution exist for the performance measures [8,9]. These analytical difficulties are due to the simultaneous presence of the repeated requests stream from the orbit and the normal stream of primary requests arrivals. Therefore, lots of attention have been paid to approximation methods, computational algorithms and tail asymptotics to estimate the performance measures [9–15].

On the other hand, the heterogeneity of customers from the point of view of customers characteristics such as the arrivals, the service and/or the retrial process distributions, is an another important problem in retrial queueing area, because models with different types of customers arise in various practical systems. For example, in cellular mobile networks, the base station channels are used by a class of fresh calls initiated in the same cell and a second class of handoff calls incoming from adjacent cells. Similarly, in modern call centers, multiple types of calls arrive at service station over different communication channels such as telephone, internet, e-mail, mobile device, etc.

However, retrial queues with multiple classes of customers (called also multiclass retrial queues) have been known to be far more difficult for mathematical analysis than models with a single class of customers (or homogeneous customers). So, explicit results for this subject are limited to some particular cases [16] and recently, sufficient stability conditions were defined for a multiserver multiclass retrial queue [17].

For the single server retrial queues with two classes of customers, a number of analytic results have been obtained [8,9,18–21]. As regards to multiserver case with two customers classes, as far as we know, there are no explicit formulae and only a few algorithmic methods are proposed using matrix geometric methods [22], matrix analytic methods [23] or computational approaches as the one we have proposed using the Colored Generalized Stochastic Petri nets formalism [24]. Recently, Kim et al. [25] studied the stability of a two-class two-server retrial queue.

The objective of this paper is to propose a new recursive algorithmic approach for the performance analysis of a multiserver retrial queue with two classes of customers: ordinary customers whose access to the service depends on the number of available servers and who join the retrial group with a certain degree of impatience when blocked; and priority customers who have access to all servers and leave definitively the system when no server is available. Hence, and in order to minimize the loss probability of priority customers, they should be given a higher priority over ordinary customers in access to the system resources (or servers). To this end, we give them the possibility to use all servers, unlike ordinary customers whose access to the service depends on a threshold on the number of available servers. Further, we assume that all servers follow the non-preemptive priority rule, which means that if one or more priority customers arrive during the service time of an ordinary customer, the current service of this non-priority customer continues and is not stopped.

Some papers considered retrial models with two customers classes and pre-emptive priority [21,23,26,27] or a non-preemptive priority in the single server case [28,29]. However, there is no work that deals with multiserver retrial queueing systems with two customers classes and non-preemptive priority. That motivates us to investigate such queueing model in this work.

The layout of the paper is given as follows: After the introduction, a detailed mathematical description of the model under study is given in Sect. 2. Then, we present our analysis approach and the details of the recursive algorithm we propose to calculate the stationary states probabilities in Sect. 3. Next, we give the formulae of the main performance measures. In Sect. 5, we discuss through numerical examples, the effect of the dedicated servers number and retrial rate on the system performances, namely the blocking probability for ordinary customers and loss probability for priority customers. Finally, we give a conclusion.

2   Mathematical Description of the Model

We consider a retrial multi-server queueing system with two classes of customers; ordinary and priority ones. The service area consists of C, $(C \geq 1)$ homogeneous servers with the same exponential service rate μ. The ordinary (priority) customers arrive in the system following a Poisson process with a mean arrival rate λ_1 (λ_2 respectively). The global arrival rate is then given by $\lambda = \lambda_1 + \lambda_2$. In order to ensure that priority customers are served prior to ordinary (non-priority) ones, our strategy consists of reserving a certain number of servers d $(1 \leq d \leq C)$, called *Dedicated Servers* only for priority class of customers. Thus, on the arrival of a priority customer, if at least one server of the C servers of the service station is idle, it will be served immediately, otherwise, it will be lost definitively, whereas an arriving ordinary customer must find at least $(d + 1)$ available servers to get service, otherwise, it joins the orbit and retry for the service later. A blocked customer in the orbit decides to retry with probability θ or give up and returns to the free state with probability $(1 - \theta)$. Note that θ is used to represent the degree of *impatience of customers*. The retrial time is exponentially distributed with rate α. All involved random variables are independent and identically distributed.

3   Recursive Analysis Algorithm

From the stochastic behavior of both two classes of customers and the servers allocation policy, the retrial system described above can be modeled by means of a two-dimensional Continuous-Time Markov Chain (CTMC) where each state is described by means of two random variables $(X(t), Y(t); t \geq 0)$. Let $X(t)$ be the number of customers being in service (which equals the number of busy servers), and $Y(t)$ the number of customers waiting in the orbit at time t. Hence, the steady state probabilities are defined by the probabilities

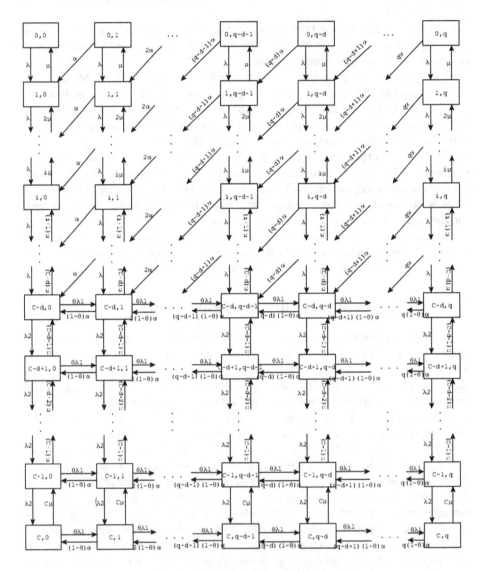

Fig. 1. State Transition Diagram

$\pi_{i,j} = Pr\{X = i, Y = j\}, i = 0, 1, \ldots, C \; j = 0, 1, \ldots, \ldots$ of having i customers in service and j customers in the orbit.

The number of states of this CTMC with $S = \{0, \ldots, C\} \times Z_+$ as state space, is infinite because the population size and the orbit capacity are supposed to be infinite. In order to obtain a finite CMTC model, we propose the truncation of the state space to $S' = \{0, \ldots, C\} \times \{0, \ldots, q\}$ with q large enough. In other terms, the probability of being in states with a number of customers in orbit greater than q is neglected.

The truncated CTMC state transition diagram is depicted in Fig. 1.

The *balance equation* describing the probability flux in and out of state (i, j) is defined by: T

$$E(i, j) : \sum_{(k,l) \in S' \backslash (i,j)} \pi_{i,j}.R_{(i,j),(k,l)} = \sum_{(k,l) \in S' \backslash (i,j)} \pi_{k,l}.R_{(k,l),(i,j)}$$

where $R_{(i,j),(k,l)}$ is the transition rate from state (i, j) to state (k, l).

We put $K_0 = \pi_{0,q}$, $K_1 = \pi_{0,(q-1)}, \ldots, K_d = \pi_{(0,q-d)}$. We first should express all probabilities as a function of K_i, $i = 0, \ldots, d$.

$$\pi_{i,j} = K_0.u_0(i, j) + K_1.u_1(i, j) + \ldots + K_d.u_d(i, j)$$

Then, we express coefficients K_i, $i = 0, \ldots, d$ as a function of K_0, and finally, we use the normalization equation, where the unique unknown is K_0,

$$\sum_{i=0}^{C} \sum_{j=0}^{q} \pi_{i,j} = 1 \tag{1}$$

to find its value.

We now proceed to explain the details of the algorithm:

Step1. Expressing all probabilities in K_i.

1. Columns q down to $q - d$.

Starting with column q, it's obvious that $\pi_{(0,q)} = K_0$, such as $u_0(0, q) = 1$, $u_1(0, q) = 0, \ldots, u_d(0, q) = 0$.

We calculate recursively, for $i = 1, \ldots, C - (d + 1)$, $\pi_{i,q}$ using the balance equation $E(i - 1, q)$. We get:

$$\pi_{1,q} = \frac{q.\alpha + \lambda}{\mu}.\pi_{0,q}$$

$$\pi_{i+1,q} = \frac{q.\alpha + \lambda + i.\mu}{(i + 1)\mu}\pi_{i,q} - \frac{\lambda}{(i + 1)\mu}.\pi_{i-1,q}$$

In the same way, we calculate for columns $(q - j)$, $j = 1, \ldots, d$, the value of $\pi_{1,q-j}$ using $E(0, q - j)$ first, then $\pi_{i+1,q-j}$ using $E(i, q - j)$, $i = 0, \ldots, C - d - 1$. We get:

$$\pi_{1,q-j} = \frac{(q - j).\alpha + \lambda}{\mu}.K_j$$

$$\pi_{i+1,q-j} = \frac{(q - j).\alpha + \lambda + i.\mu}{(i + 1)\mu}\pi_{i,q-j} - \frac{\lambda}{(i + 1)\mu}\pi_{i-1,q-j} - \frac{(q - j + 1)\alpha}{(i + 1)\mu}\pi_{i-1,q-j+1}$$

Then, inside the line $(C - d + 1)$, columns from $j = (q - d + 1)$ to $j = (q - 1)$ can be calculated. Actually, for each probability $\pi_{C-d+1,j}$, we use the balance equation $E(C - d, j)$:

$$\pi_{C-d+1,j}.[(C - d + 1)\mu] = [(C - d)\mu + \theta\lambda_1 + \lambda_2 + j(1 - \theta)\alpha].\pi_{C-d,j} - \lambda.\pi_{C-d-1,j}$$

$$- [(1 - \theta)(j + 1)\alpha].\pi_{C-d,j+1} - \theta\lambda_1.\pi_{C-d,j-1} - [(j + 1)\alpha].\pi_{C-d-1,j+1}$$

And for $\pi_{C-d+1,q}$, we use $E(C-d,q)$, we have:

$$\pi_{C-d+1,q}.[(C-d+1)\mu] =$$

$$[(C-d)\mu + \lambda_2 + (1-\theta)q\alpha].\pi_{C-d,q} - \lambda.\pi_{C-d-1,q} - \theta\lambda_1.\pi_{C-d,q-1}$$

For the rest of lines, i.e. from $i = C - d + 2$ to $i = C$, only columns $j = i + q - C, \ldots, q$ can be deduced for the moment, they are calculated from balance equations $E(i-1,j)$:
We have for $j = i + q - C$ to $j = q - 1$:

$$\pi_{i,j}.(i.\mu) = [(i-1).\mu + (1-\theta).j.\alpha + \lambda_2 + \theta.\lambda_1].\pi_{(i-1),j}$$

$$-\lambda_2.\pi_{(i-2),j} - (1-\theta).(j+1).\alpha.\pi_{(i-1),(j+1)} - \theta.\lambda_1.\pi_{(i-1),(j-1)}$$

And when $j = q$:

$$\pi_{i,q}.(i.\mu) = [(i-1).\mu + (1-\theta).q.\alpha + \lambda_2].\pi_{(i-1),q} - \lambda_2.\pi_{(i-2),q} - \theta.\lambda_1.\pi_{(i-1),(q-1)}$$

From $E(C,q)$, we obtain the value of $\pi_{C,q-1}$ as follows:

$$\pi_{C,q-1}.\theta.\lambda_1 = [C.\mu + (1-\theta).q.\alpha].\pi_{C,q} - \lambda_2.\pi_{(C-1),q}$$

Now, in order to have the rest of columns, we proceed like this. For each $j = q - 2$ to $q - d$, we obtain first $\pi_{C,j}$ using $E(C, j+1)$:

$$\pi_{C,j}.\theta.\lambda_1 = [(1-\theta).(j+1).\alpha + C.\mu + \theta.\lambda_1].\pi_{C,(j+1)}$$

$$-(1-\theta).(j+2).\alpha.\pi_{C,(j+2)} - \lambda_2.\pi_{(C-1),(j+1)}$$

Then, we obtain the other lines $i = C - 1$ to $i = C + 1 + j - q$, thanks to $E(i, j+1)$:

$$\pi_{i,j}.\theta.\lambda_1 = [(1-\theta).(j+1).\alpha + i.\mu + \theta.\lambda_1 + \lambda_2].\pi_{i,(j+1)}$$

$$-(1-\theta).(j+2).\alpha.\pi_{i,(j+2)} - \lambda_2.\pi_{(i-1),(j+1)} - (i+1).\mu.\pi_{(i+1),(j+1)}$$

Up to now, we have expressed all probabilities from column $j = q$ to $j = q-d$, in $K_0, \ldots, K_d$.

2. Column $(q - d - 1)$ down to 1.

In a similar way as explained above, by invoking $E(i-1,j)$, for each $\pi_{i,j}$, we find:

$$\pi_{1,j} = \frac{j.\alpha + \lambda}{\mu}.\pi_{0,j}$$

$$\pi_{i+1,j} = \frac{j.\alpha + \lambda + i.\mu}{(i+1)\mu}\pi_{i,j} - \frac{\lambda}{(i+1)\mu}\pi_{i-1,j} - \frac{(j+1)\alpha}{(i+1)\mu}\pi_{i-1,j+1}$$

In particular, when $i = C - d - 1$, we can find numbers $v_0(C - d, j), v_1(C - d, j), \ldots, v_{d+1}(C - d, j)$, such that:

$$\pi_{(C-d),j} = v_0(C-d,j).K_0 + \ldots + v_d(C-d,j).K_d + v_{(d+1)}(C-d,j).\pi_{0,j} \quad (2)$$

On the other hand, by invoking $E(C-d, j+1)$, we get:

$$\pi_{(C-d),j}.\theta.\lambda_1 = [(C-d).\mu + \theta.\lambda_1 + \lambda_2 + (1-\theta).(j+1).\alpha].\pi_{(C-d),(j+1)} - \lambda.\pi_{(C-d-1),(j+1)}$$

$$-(j+2).\alpha.\pi_{(C-d-1,j+2)} - (1-\theta).(j+2).\alpha.\pi_{(C-d),(j+2)} - (C-d+1).\mu.\pi_{(C-d+1),(j+1)}$$

which implies that we can find explicitly numbers $u_0(C-d,j), u_1(C-d,j), \ldots, u_d(C-d,j)$, such that:

$$\pi_{(C-d),j} = u_0(C-d,j).K_0 + u_1(C-d,j).K_1 + \ldots + u_d(C-d,j).K_d \quad (3)$$

From Eqs. (2) and (3), we can deduce $\pi_{0,j}$ value,

$$\pi_{0,j} = \frac{\sum_{k=0}^{d}[u_k(C-d,j) - v_k(C-d,j)].K_k}{v_{(d+1)}(C-d,j)}$$

Thus, we calculate again $\pi_{i,j}$, $i = 1, \ldots, C-d$, in $K_0, K_1, \ldots, K_d$ only, we get for $k = 0, \ldots, d$:

$$u_k(i,j) = v_k(i,j) + \frac{u_k(C-d,j) - v_k(C-d,j)}{v_{(d+1)}(C-d,j)}.v_{(d+1)}(i,j)$$

After that, equations for $\pi_{i,j}$, such that $i = C-d+1, \ldots, C$ can be easily derived from $E(i, j+1)$.

Step2. Expressing coefficients $K_i, i = 1, \ldots, d$ in K_0.

Let's consider the balance equation $E(C,0)$:

$$\pi_{C,0}.(C.\mu + \theta.\lambda_1) = \pi_{(C-1),0}.\lambda_2 + \pi_{C,1}.(1-\theta).\alpha$$

Keeping in mind that both $\pi_{C,0}$, $\pi_{(C-1),0}$ and $\pi_{C,1}$ can be written as a linear combination of $K_0, \ldots, K_d$, Eq. (3) is equivalent to:

$$\sum_{k=0}^{d} u_k(C,0).K_k.(C.\mu + \theta.\lambda_1) = \sum_{k=0}^{d} u_k(C-1,0).K_k.\lambda_2 + \sum_{k=0}^{d} u_k(C,1).K_k.(1-\theta).\alpha$$

It's a question of a simple algebra to extract K_d in $K_{(d-1)}, \ldots, K_0$. In the same way, we consider $E(i,0)$, $i = C-1, \ldots, C-d+1$, to have K_x in $K_{x-1}, \ldots, K_0$, $x = d-1, \ldots, 1$.

Step3. Finding K_0.

Finally, we solve the normalization Eq. (1), in order to extract the value of K_0 which is the unique unknown.

4   Performance Measures

Once all the stationary probabilities are determined thanks to the above algorithm, several performance measures can be given so as the efficiency of our system can be judged. The most significant performance indices are as follows:

– Mean number of busy servers:

$$N_{Busy} = \sum_{i=0}^{C}\sum_{j=0}^{q} i.\pi_{i,j}$$

– Mean number of customers in the orbit:

$$N_{Orbit} = \sum_{i=0}^{C}\sum_{j=0}^{q} j.\pi_{i,j}$$

– Mean number of customers in the system:

$$N = N_{Busy} + N_{Orbit} = \sum_{i=0}^{C}\sum_{j=0}^{q}(i+j).\pi_{i,j}$$

– Mean rate of ordinary customers served at the first attempt:

$$\bar{\lambda}_{FS} = \lambda_1. \sum_{i=0}^{C-(d+1)} \sum_{j=0}^{q} .\pi_{i,j}$$

– Mean rate of blocked ordinary customers:

$$\bar{\lambda}_{FU} = \lambda_1.\theta. \sum_{i=C-d}^{C} \sum_{j=0}^{q} .\pi_{i,j}$$

– Mean rate of blocked ordinary customers leaving the system without being served:

$$\bar{\lambda}_{FB} = \lambda_1.(1-\theta). \sum_{i=C-d}^{C} \sum_{j=0}^{q} \pi_{i,j}$$

– Effective mean ordinary customers arrival rate:

$$\bar{\lambda}_F = \bar{\lambda}_{FS} + \bar{\lambda}_{FU} + \bar{\lambda}_{FB}$$

– Mean rate of retrials served at the first attempt:

$$\bar{\alpha}_{RS} = \alpha. \sum_{i=0}^{C-(d+1)} \sum_{j=0}^{q} j.\pi_{i,j}$$

- Mean rate of blocked retrials:

$$\bar{\alpha}_{RU} = \alpha.\theta. \sum_{i=C-d}^{C} \sum_{j=0}^{q} j.\pi_{i,j}$$

- Mean rate of blocked retrials leaving the system without being served:

$$\bar{\alpha}_{RB} = \alpha.(1-\theta). \sum_{i=C-d}^{C} \sum_{j=0}^{q} j.\pi_{i,j}$$

- Effective mean retrial rate:

$$\bar{\alpha} = \bar{\alpha}_{RS} + \bar{\alpha}_{RU} + \bar{\alpha}_{RB}$$

- Mean rate of priority customers being served:

$$\bar{\lambda}_{HS} = \lambda_2. \sum_{i=0}^{C-1} \sum_{j=0}^{q} \pi_{i,j}$$

- Mean rate of lost priority customers:

$$\bar{\lambda}_{HB} = \lambda_2. \sum_{j=0}^{q} \pi_{C,j}$$

- Effective mean priority customers arrival rate:

$$\bar{\lambda}_{H} = \bar{\lambda}_{HS} + \bar{\lambda}_{HB}$$

- Mean service rate:

$$\bar{\mu} = \mu. \sum_{i=1}^{C} \sum_{j=0}^{q} i.\pi_{i,j} = \mu.N_{Busy}$$

- Blocking probability of ordinary customers:

$$P_{BF} = \frac{\bar{\lambda}_{FU} + \bar{\lambda}_{FB}}{\bar{\lambda}_{F}}$$

- Blocking probability of retrial customers:

$$P_{BR} = \frac{\bar{\alpha}_{RU} + \bar{\alpha}_{RB}}{\bar{\alpha}}$$

- Loss probability (of priority customers):

$$P_{BH} = \frac{\bar{\lambda}_{HB}}{\bar{\lambda}_{H}}$$

- Mean number of attempts per customer:

$$\eta = \frac{\bar{\alpha} + \bar{\lambda}_F}{\bar{\lambda}_F} = 1 + \frac{\bar{\alpha}}{\bar{\lambda}_F}$$

- Availability of at least s servers:

$$A_s = \sum_{i=0}^{C-s} \sum_{j=0}^{q} \pi_{i,j}$$

- Mean waiting time of an ordinary customer:

$$\bar{W}_F = \frac{N_{Orbit}}{\bar{\lambda}_1}$$

5   Numerical Results

In this section, we examine the impact that have some system parameters like the number of dedicated servers, degree of persistence and the arrival and service rates on the system performance, namely the blocking and loss probability.

Table 1. System parameters.

C	$1/\mu$	α/μ	λ_1/λ_2
15	120 s	20	24

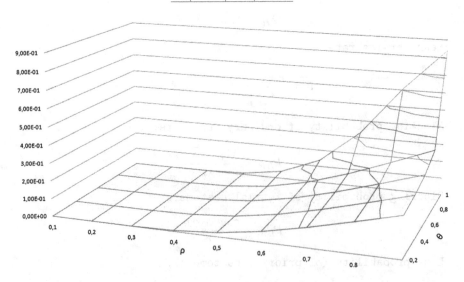

Fig. 2. Influence of the offered traffic and the degree of persistence on the blocking probability (with $d = 2$).

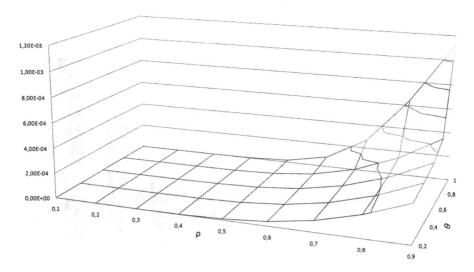

Fig. 3. Influence of the offered traffic and the degree of persistence on the loss probability (with $d = 2$).

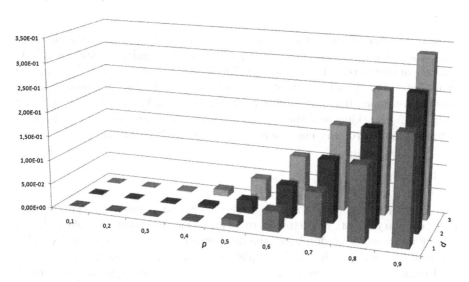

Fig. 4. Influence of the offered traffic and the number of dedicated servers on the blocking probability (with $\theta = 0.6$).

We developed a $C\#$ code in order to implement the above algorithm. In that follows, unless otherwise stated, the parameter set depicted in Table 1 is used for our experiments. The offered load is defined by $\rho = \lambda/(C.\mu)$.

The effect of the traffic load ρ and the persistence degree θ on the blocking and the loss probability is shown in Figs. 2 and 3 respectively (with $d = 2$). We can note that the increase in parameters ρ and θ affects negatively both

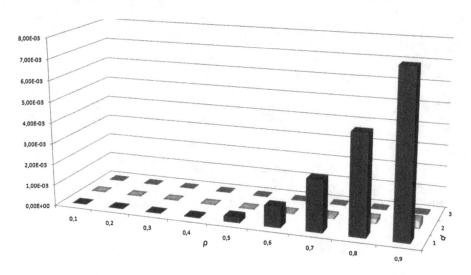

Fig. 5. Influence of the offered traffic and the number of dedicated servers on the loss probability (with $\theta = 0.6$).

of the two probabilities, but loss probability is always better than the blocking one. This is due to the number of servers dedicated for priority customers. The influence of this latter is the purpose of Figs. 4 and 5. As expected, increasing the number of dedicated servers can significantly improve the blocking probability of priority customers. It is just perfect ($\simeq 0$) when $d = 3$. We observe the opposite effect on the blocking probability, when more servers are reserved to priority customers, more ordinary customers are blocked at their arrival.

6   Conclusion

A multiserver retrial queue with two classes of customers was investigated in this paper; ordinary impatient customers retry for service when they are blocked, and priority customers who don't join the orbit when all servers are occupied. This is a realistic model as it can be applicable to various situations encountered in telecommunication networks, call centers, manufacturing systems, etc., which deal with two types of customers under certain priority rule. In order to minimize the loss probability of priority customers, they should be given a higher priority over ordinary customers in access to the system servers. Our proposition was to reserve some servers to be used only by priority customers. The analysis of the model was performed using a bi-dimensional Time Continuous Markov chain, and an efficient recursive algorithm was proposed and implemented in order to calculate the steady state probability distribution. Moreover, the formulae of several performance measures were developed. We showed via numerical examples that dedicated servers technique improves the system performance, mainly the loss probability, but at the expense of ordinary customers.

References

1. Kim, C., Klimenok, V.I., Dudin, A.N.: Analysis and optimization of Guard Channel Policy in cellular mobile networks with account of retrials. Comput. Oper. Res. **43**, 181–190 (2014)
2. Gharbi, N., Nemmouchi, B., Mokdad, L., Ben-Othmane, J.: The impact of breakdowns disciplines and repeated attempts on performances of small cell networks. J. Comput. Sci. **5**, 633–644 (2014)
3. Gharbi, N., Charabi, L., Mokdad, L.: Performance evaluation of heterogeneous servers allocation disciplines in networks with retrials. In: The IEEE 17th International Conference on High Performance Computing and Communications (HPCC 2015), New York, USA, 24–26, August (2015)
4. Gharbi, N.: Using GSPNs for performance analysis of a new admission control strategy with retrials and guard channels. In: The 3rd International Conference on Mobile and Wireless Technology 2016 (ICMWT 2016), Korea (2016)
5. Charabi, L., Gharbi, N., Ben-Othman, J., Mokdad, L.: Call admission control in small cell networks with retrials and guard channels. In: Proceedings of The IEEE Global Communications Conference 2016 (GLOBECOM 2016), USA (2016)
6. Wüchner, P., Sztrik, J., Meer, H.: Modeling wireless sensor networks using finite-source retrial queues with unreliable orbit. In: Hummel, K.A., Hlavacs, H., Gansterer, W. (eds.) PERFORM 2010. LNCS, vol. 6821, pp. 73–86. Springer, Heidelberg (2011). doi:10.1007/978-3-642-25575-5_7
7. Falin, G.I., Templeton, J.G.C.: Retrial Queues. Chapman and Hall, London (1997)
8. Artalejo, J.R., Gmez-Corral, A.: Retrial Queueing Systems: A Computational Approach. Springer, Berlin (2008)
9. Kim, J., Kim, B.: A survey of retrial queueing systems. Ann. Oper. Res. 1–34 (2015)
10. Do, T.V., Do, N.H., Zhang, J.: An enhanced algorithm to solve multiserver retrial queueing systems with impatient customers. Comput. Ind. Eng. **65**(4), 719–728 (2013)
11. Dudin, A.N., Dudina, O.S.: Analysis of multiserver retrial queueing system with varying capacity and parameters. Math. Probl. Eng. 2015 (2015)
12. Kim, B., Kim, J.: Exact tail asymptotics for the $M/M/m$ retrial queue with non-persistent customers. Oper. Res. Lett. **40**, 537–540 (2012)
13. Kim, J., Kim, J., Kim, B.: Tail asymptotics of the queue size distribution in the M/M/m retrial queue. J. Comput. Appl. Math. **236**, 3445–3460 (2012)
14. Kim, J., Kim, J.: Waiting time distribution in the M/M/m retrial queue. Bull. Korean Math. Soc. **50**, 1659–1671 (2013)
15. Phung-Duc, T., Masuyama, H., Kasahara, S., Takahashi, Y.: A matrix continued fraction approach to multiserver retrial queues. Ann. Oper. Res. **202**, 161–183 (2013)
16. Shin, Y.W., Moon, D.H.: M/M/c retrial queue with multiclass of customers. Method. Comput. Appl. Probab. **16**, 931–949 (2014)
17. Avrachenkov, K., Morozov, E., Steyaert, B.: Sufficient stability conditions for multi-class constant retrial rate systems. Queueing Syst. **82**(1), 149–171 (2016)
18. Artalejo, J.R., Phung-Duc, T.: Single server retrial queues with two way communication. Appl. Math. Model. **37**, 1811–1822 (2013)
19. Avrachenkov, K., Nain, P., Yechiali, U.: A retrial system with two input streams and two orbit queues. Queueing Syst. **77**(1), 1–31 (2014)

20. Choi, B.D., Chang, Y.: Single server retrial queues with priority calls. Math. Comput. Model. **30**(3–4), 7–32 (1999)
21. Gao, S.: A preemptive priority retrial queue with two classes of customers and general retrial times. Oper. Res. **15**(2), 233–251 (2015)
22. Choi, B.D., Chang, Y.: MAP1, MAP2/M/c retrial queue with the retrial group of finite capacity and geometric loss. Math. Comput. Model. **30**, 99–113 (1999)
23. Kumar, M.S., Chakravarthy, S.R., Arumuganathan, R.: Preemptive resume priority retrial queue with two classes of MAP arrivals. Appl. Math. Sci. **7**(52), 2569–2589 (2013)
24. Gharbi, N., Dutheillet, C., Ioualalen, M.: Colored stochastic petri nets for modelling and analysis of multiclass retrial systems. Math. Comput. Model. **49**, 1436–1448 (2009)
25. Kim, B., Kim, J.: Stability of a two-class two-server retrial queueing system. Perform. Eval. **88–89**, 1–17 (2015)
26. Boutarfa, L., Djellab, N.: On the performance of the M1, M2/G1, G2/1 retrial queue with pre-emptive resume policy. Yugoslav J. Oper. Res. **25**(1), 153–164 (2015)
27. Dimitriou, I.: A preemptive resume priority retrial queue with state dependent arrivals, unreliable server and negative customers. TOP **21**(3), 542–571 (2013)
28. Ayyapan, G., Muthu Ganapathi Subramanian, A., Sekar, G.: M/M/1 retrial queueing system with loss and feedback under non-pre-emptive priority service by matrix geometric method. Appl. Math. Sci. **4**(48), 2379–2389 (2010)
29. Madan, K.C.: A non-preemptive priority queueing system with a single server serving two queues M/G/1 and M/D/1 with optional server vacations based on exhaustive service of the priority units. Appl. Math. **2**, 791–799 (2011)

Markovian Queue with Garbage Collection

Illés Horváth[1]($\boxtimes$), István Finta[2], Ferenc Kovács[2], András Mészáros[3],
Roland Molontay[4], and Krisztián Varga[2]

[1] MTA-BME Information Systems Research Group, Budapest, Hungary
pollux@math.bme.hu
[2] Nokia, Bell Labs, Budapest, Hungary
[3] Department of Networked Systems and Services,
Budapest University of Technology and Economics, Budapest, Hungary
[4] Department of Stochastics,
Budapest University of Technology and Economics, Budapest, Hungary

Abstract. Garbage collection is a fundamental component of memory management in several software frameworks. We present a general two-dimensional Markovian model of a queue with garbage collection where the input process is Markov-modulated and the memory consumption can be modeled with discretisation. We derive important performance measures (also including garbage collection-related measures like mean garbage collection cycle length). The model is validated via measurements from a real-life data processing pipeline.

Keywords: Memory management · Garbage collection · Stochastic modelling · Markovian modelling

1 Introduction

Some of the most popular languages such as Java and C# require efficient memory management including garbage collection (GC): the automated process of identifying and recovering the storage space that is occupied by objects that are no longer required. The physical memory is a limited resource so designing more sophisticated garbage collectors and providing a theoretical framework are of particular research interest.

A huge number of different garbage collection techniques have been proposed throughout the years. For surveys and evaluation of garbage collection algorithms we refer the reader to [11,22,27]. More recent GC techniques include Garbage First for multi-processors with large memory [9], Metronome, a real-time GC integrated with the scheduling system [3], MMTk, a memory management toolkit for Java [5], the concurrent-copy collector, a real time garbage collector for Java [23], and FeGC, an efficient GC scheme for flash memory based storage systems [13]. Virtual machine garbage collection optimization was addressed in [4]; the model shares the main idea with the present paper (discretisation of the memory), with an overall simpler, essentially 1-dimensional model with focus on optimising parameters for garbage collection.

© Springer International Publishing AG 2017
N. Thomas and M. Forshaw (Eds.): ASMTA 2017, LNCS 10378, pp. 109–124, 2017.
DOI: 10.1007/978-3-319-61428-1_8

A more recent tendency has been to consider formal models therefore provide a rigorous method to characterize GC algorithms and analyze their performance independently of the programming system. In particular, analytical modelling of garbage collection algorithms in flash-based solid-state drive (SSD) systems has received significant research interest [8,15,25,26,28].

The majority of the analytical studies on garbage collection process have focused on the following specific algorithms: greedy GC, FIFO GC, Windowed GC, d-choices GC. For more details, see [25,28].

An important issue regarding garbage collection in SSD systems is the so-called write amplification phenomenon. For more details, see [28].

A number of analytical frameworks have been proposed by the abstraction of the block state space and the stochastic modelling of the selection process. In [8] a Markov chain model is provided to characterize the performance of SSD operation for uniformly distributed random small user writes and considering the greedy scheme. They find that write amplification increases as the system occupancy increases as the number of pages per block increases but decreases as the number of block increases. In [15] a Markov chain model is employed to capture the dynamics of large-scale SSDs, and mean-field theory is applied to derive the asymptotic steady state, the performance/durability trade-off of GC algorithms is analyzed. Yang et al. also apply mean field analysis and show that the system dynamics can be represented by a system of ordinary differential equations and the steady state of the write amplification can be predicted for a class of GC algorithms (including d-choices) [28].

Another modelling approach is providing a theoretical framework of distributed garbage collection [7,16,21]. The increasing use of distributed systems implies that distributed garbage collectors should be considered. A formal model of distributed garbage collection is Surf [7] that can describe a wide range of GCs and is amenable to rigorous analysis.

In our work, we focus on a Markovian approach that models of the effect of garbage collection on memory management. We present the model in two steps. In Sect. 2.1, we present a 2-dimensional Markov-modulated fluid description of the model. The fluid approach is easy to define but difficult to solve analytically. Then we present the corresponding Markovian model in Sect. 2.2, which is essentially a discretisation of the memory level. The Markovian model can be solved efficiently numerically, with the analysis and performance measures derived in Sect. 3.

Section 4 contains an application to an actual data processing system. The model of Sect. 2.2 is then validated by comparison to performance measurements of the actual system.

2   Queue with Garbage Collection

In the model, data arrives at a server and is stored in the memory. When it is processed, it does not flush (empty) immediately from the memory, but is only flushed when the memory reaches a certain level.

Memory level is described by two variables: in-use memory (V) and junk memory (U). In-use memory contains all data that has not been processed yet (that is, the queue), while junk memory contains data that has been processed since the last GC period. Data processing may generate extra memory usage; we assume that this extra memory usage is proportional to the size of the data with multiplicative constant C (that is, processing 1 byte of data creates a total of C bytes of memory usage in addition to the original 1 byte). We also make the assumption that the service time of data is proportional to the amount of data; this assumption means that V is proportional to the service queue length. These assumptions typically hold for systems with relatively simple processing.

When the total memory $U+V$ reaches a certain level M, GC turns on. During GC, the junk memory flushes at a fixed rate g, but data may keep arriving (and stored entirely in in-use memory). For simplicity, we assume that there is no service during GC. When GC finishes, service is resumed.

We assume that arrivals are Markov-modulated with a finite state space S and generator Q. The arrival process itself is denoted by $X(t)$. The arrival rate in state $i \in S$ is r_i, and the service rate is constant s.

In Sect. 2.1, we present a fluid approach to model the memory level. While the model definition is relatively straightforward and tidy from the behaviour of the system, it leads to a 2-dimensional fluid queue with special behaviour on the boundaries.

2.1   Fluid Description

Fluid modelling approach is an efficient way of describing and analyzing a wide range of real systems for domains as diverse as job scheduling [19] and battery life [12]. An overview of the basic concepts of fluid models with the potential usage in performance analysis can be found in [10].

A fluid description of the queue is obtained when data is assumed to be continuous; in this case, U (junk memory level) and V (queue) are fluid variables governed by the arrival process (a continuous time Markov chain) and the switch between service and GC modes.

The behaviour of the system is governed by the equations

$$
\left.\begin{array}{l} dU(t)/dt = Cs \\ dV(t)/dt = r_{X(t)} - s \end{array}\right\} \text{ if } V(t) > 0 \text{ during service}
$$

$$
\left.\begin{array}{l} dU(t)/dt = Cs \\ dV(t)/dt = r_{X(t)} - s \end{array}\right\} \text{ if } V(t) = 0 \text{ and } r_{X(t)} > s \text{ during service}
$$

$$
\left.\begin{array}{l} dU(t)/dt = Cr_{X(t)} \\ dV(t)/dt = 0 \end{array}\right\} \text{ if } V(t) = 0 \text{ and } r_{X(t)} < s \text{ during service} \qquad (1)
$$

$$
\left.\begin{array}{l} dU(t)/dt = -g \\ dV(t)/dt = r_{X(t)} \end{array}\right\} \text{ during GC}
$$

and the forced transitions:

- when $U(t) + V(t)$ reaches M during service, we switch to GC mode;
- when $U(t)$ reaches 0 during GC, we switch to service mode.

In (1), $V(t) = 0$ corresponds to no queue; if $r_{X(t)<s}$, all incoming data is processed immediately, while if $r_{X(t)>s}$, the queue starts growing. As long as $V(t) > 0$, the server is working at a full service rate. During garbage collection, there is no service, so all incoming data goes in the queue.

The above system is difficult to solve analytically. For 2-dimensional fluid queues, very few results available. Instead, we present a discretised Markovian version of the model in Sect. 2.2 where U and V are both discretised; stationary analysis of the Markovian model is carried out in Sect. 3. A detailed analysis of the original fluid model is subject to further research.

The Markov model of Sect. 2.2 is applied to a data processing application in Sect. 4 with the performance measures predicted by the model compared with measurements from the real system. We note that only some of the measures derived in Sect. 3 are measured in the application. We nevertheless included these and other measures as well in Sect. 3, with possible different future applications in mind.

2.2   Markovian Description

Markovian queuing theory is a well-established topic with diversified domains of application. For a detailed introduction to queuing theory with computer science and telecommunication applications we refer the reader to [6,18,24].

In this approach, we replace the fluid queues U and V by a discrete memory level to obtain a Markovian model with a discrete state space. We note that we allow U and V to be discretised with different granularity; assume the possible values of U are divided into N_U different sections, while the possible values of V are divided into N_V different sections. The reason to allow a different granularity lies in the fact that the behaviour of the system depends highly on whether $V = 0$ or $V > 0$: as long as $V > 0$ (there is a queue), the system will work at full capacity. Thus it makes sense to select N_V relatively high in order to be able to identify $V > 0$ more precisely. Since the exact value of U is less relevant in the behaviour of the system (apart from the total memory reaching M), the granularity of U may be allowed to be less fine. For simplicity, we assume N_V is an integer multiple of N_U.

The maximal possible memory level M corresponds to a full memory, while the value 0 corresponds to an empty memory that contains no data. In correspondence with this, in the Markovian description U and V refer to the level of junk memory and in-use memory, respectively, and can only take (non-negative) integer values such that $0 \leq U \leq N_U$ and $0 \leq V < N_V$. Note that in applications, the memory level corresponds to memory used exclusively for the processing of data; memory usage by other system processes is not included.

Altogether, the following parameters define the system:

- M, the value of the memory cap;
- the state space $\mathcal{S}$, the generator Q and the rate vector $\{r_i : i \in \mathcal{S}\}$ define the arrival process;
- s is the service rate;

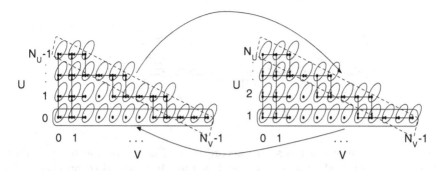

Fig. 1. State space (left-hand side: service, right-hand side: GC)

- g is the rate of garbage collection;
- C is the ratio of memory usage generated during processing compared to the size of the data;
- N_U and N_V describe the granularity of the memory.

We obtain a finite 4-dimensional state space Ω with

- dimension 1 representing the arrival process, and
- dimension 2 representing the value of V (in-use memory),
- dimension 3 representing the value of U (junk-memory),
- dimension 4 representing GC or service mode.

We assume GC starts immediately when $U/N_U + V/N_V$ reaches 1 during service, and service restarts immediately when U reaches 0 during GC. Thus, for the service states $U/N_U + V/N_V < 1$ holds, which can be depicted as a triangular shaped array. For the GC states, the value of U is essentially higher by 1: $1 \leq U \leq N_U$, and service restarts immediately when U reaches 0. The state space is depicted in Fig. 1. Dimension 1 is depicted only in small bubbles; dimensions 2 and 3 are represented by the large triangular arrays, and dimension 4 only has size 2, which is depicted as the two triangles on the left and right. Altogether,

$$\Omega = \{(i, j, k, \text{service}) : i \in S, j \geq 0, k \geq 0, j/N_V + k/N_U < 1\} \cup$$
$$\{(i, j, k, \text{GC}) : i \in S, j \geq 1, k \geq 0, (j-1)/N_V + k/N_U < 1\}. \quad (2)$$

In the following notations, i denotes the state of the background process, j denotes the value of V, k denotes the value of U, and l denotes the service mode (either service or GC). The type of transitions possible from a state depend slightly on where the state is situated within the two triangles; three main types of transitions are present: transitions corresponding to (1) the changes of the background process, (2) arrivals, and (3) service. Arrivals are suppressed when memory is full. Service increases the junk memory and decreases the queue (decreasing is suppressed when the queue is empty). Also, the forced transitions are present at the diagonal border of the service triangle and the bottom row of the GC triangle.

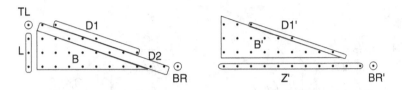

Fig. 2. States grouping

For clarity, we group the states according to Fig. 2. For the service states (left triangle), TL is the top left corner, L is the left border (except TL). BR is the bottom right corner. D1 is the rightmost nodes in the diagonal (except BR), D2 is the rest of the diagonal (the "uppermost" states except D1 and TL). The remaining states are grouped together in B, the "bulk" of the service states. For the GC states, D1' and BR' are the counterparts of D1 and BR, Z' is the bottom row (the states from which service may resume) and B' is the bulk of the GC states. We omit a more formal definition of the groups. Note that the groups are understood within the 3-dimensional subspace of (U, V) and the service type.

We have the following types of transitions.

The background process may change at any state regardless of memory levels or service mode:

– for any (i, j, k, l), we may transition from (i, j, k, l) to (i', j, k, l) according to the generator Q of the arrival process;

For $(j, k, l) \in B$, that is, the bulk of the service states, we have different types of transitions depending on whether $r_i < s$ or $r_i > s$. If $r_i < s$, we have the following transitions:

– from $(i, j, k, \text{service})$ to $(i, j, k + 1, \text{service})$ with rate $N_U \cdot C \cdot s/M$,
– from $(i, j, k, \text{service})$ to $(i, j - 1, k, \text{service})$ with rate $N_V (s - r_i)/M$;

while if $r_i > s$, we have the transitions:

– from $(i, j, k, \text{service})$ to $(i, j, k + 1, \text{service})$ with rate $N_U \cdot C \cdot s/M$,
– from $(i, j, k, \text{service})$ to $(i, j + 1, k, \text{service})$ with rate $N_V (r_i - s)/M$.

The above values ensure that the horizontal and vertical drifts in the Markovian model are in correspondence with the fluid model (1); the factors N_U, N_V and $1/M$ are due to the discretisation.

For $(0, k, l) \in L$, that is, the queue is empty during a service period, the types of transitions depend on whether $r_i > s$ or $r_i < s$. If $r_i < s$, we have the following transitions:

– from $(i, 0, k, \text{service})$ to $(i, 0, k + 1, \text{service})$ with rate $N_U \cdot C \cdot r_i/M$,

while if $r_i > s$, we have the following transitions:

– from $(i, 0, k, \text{service})$ to $(i, 0, k + 1, \text{service})$ with rate $N_U \cdot C \cdot s/M$,
– from $(i, 0, k, \text{service})$ to $(i, 1, k, \text{service})$ with rate $N_V (r_i - s)/M$;

For $(j, k, l) \in D_2$, $r_i > s$, we also have some transitions corresponding to the forced transitions from service to GC. If $r_i < s$, we have the transitions:

- from $(i, j, k, \text{service})$ to $(i, j, k + 1, \text{GC})$ with rate $N_U \cdot C \cdot s/M$,
- from $(i, j, k, \text{service})$ to $(i, j - 1, k, \text{service})$ with rate $N_V(s - r_i)/M$;

while if $r_i > s$, we have the transitions:

- from $(i, j, k, \text{service})$ to $(i, j, k + 1, \text{GC})$ with rate $N_U \cdot C \cdot s/M$,
- from $(i, j, k, \text{service})$ to $(i, j + 1, k, \text{service})$ with rate $N_V(r_i - s)/M$.

$(j, k, l) \in D_1$ is similar; if $r_i < s$, we have the transitions:

- from $(i, j, k, \text{service})$ to $(i, j, k + 1, \text{GC})$ with rate $N_U \cdot C \cdot s/M$,
- from $(i, j, k, \text{service})$ to $(i, j - 1, k, \text{service})$ with rate $N_V(s - r_i)/M$;

while if $r_i > s$, we have the transitions:

- from $(i, j, k, \text{service})$ to $(i, j, k + 1, \text{GC})$ with rate $N_U \cdot C \cdot s/M$,
- from $(i, j, k, \text{service})$ to $(i, j + 1, k, \text{GC})$ with rate $N_V(r_i - s)/M$ (note that $N_U \leq N_V$ ensures that $(i, j + 1, k, \text{GC}) \in \Omega$).

For $(j, k, l) \in TL$ (which contains a single element, $(j, k, l) = (0, N_U - 1, \text{service})$), for $r_i < s$, we have

- from $(i, 0, N_U - 1, \text{service})$ to $(i, 0, N_U, \text{GC})$ with rate $N_U \cdot r_i/M$;

and for $r_i > s$, we have

- from $(i, 0, N_U - 1, \text{service})$ to $(i, 0, N_U, \text{GC})$ with rate $N_U \cdot S \cdot s/M$,
- from $(i, 0, N_U - 1, \text{service})$ to $(i, 1, N_U - 1, \text{service})$ with rate $N_V(r_i - s)/M$.

For $(j, k, l) \in BR$ (which is again a single element, $(j, k, l) = (N_V - 1, 0, \text{service})$), for $r_i < s$ we have

- from $(i, N_V - 1, 0, \text{service})$ to $(i, N_V - 1, 1, \text{GC})$ with rate $N_U \cdot C \cdot s/M$,
- from $(i, N_V - 1, 0, \text{service})$ to $(i, N_V - 2, 0, \text{service})$ with rate $N_V(s - r_i)/M$,

while for $r_i > s$, we have the transitions:

- from $(i, N_V - 1, 0, \text{service})$ to $(i, N_V - 1, 1, \text{GC})$ with rate $N_U \cdot C \cdot s/M$,

and the transition increasing V is suppressed (this corresponds to data loss in the system).

For $(j, k, l) \in B'$, that is, the bulk of the GC states, we have the following transitions:

- from (i, j, k, GC) to $(i, j, k - 1, \text{GC})$ with rate $N_U \cdot g/M$,
- from (i, j, k, GC) to $(i, j + 1, k, \text{GC})$ with rate $N_V \cdot r_i/M$;

For $(j, k, l) \in D_1'$, we have the following transitions:

- from (i, j, k, GC) to $(i, j, k - 1, \text{GC})$ with rate $N_U \cdot g/M$, and the transition only increasing V is suppressed; this corresponds to data loss in the system.

For $(j, k, l) \in Z'$, we have the transitions:

- from $(i, j, 1, \text{GC})$ to $(i, j, 0, \text{service})$ with rate $N_U \cdot g/M$;
- from (i, j, k, GC) to $(i, j + 1, k, \text{GC})$ with rate $N_V \cdot r_i/M$.

For $(j, k, l) \in BR'$, we have the following transitions:

- from $(i, N_V - 1, 1, \text{GC})$ to $(i, N_V - 1, 0, \text{service})$ with rate $N_U \cdot g/M$

and transitions increasing V are suppressed; these contribute to data loss.

The collection of the above transitions define a CTMC on the state space Ω.

3   Stationary Analysis

From the stationary analysis of such a system, it is possible to derive the following parameters:

- distribution and mean of memory level (both in-use and junk memory);
- mean period length (of an entire service + GC cycle, or the two separately);
- mean time spent with GC;
- mean utilisation (along with the ratio of CPU usage spent on GC and service);
- effective long-term rate of service;
- mean loss ratio and mean loss rate;
- average response time (in Sect. 3.1).

We calculate them as follows. If $v_{st}(i, j, k, l)$ denotes the stationary distribution of the system, then the mean memory levels can be calculated as follows:

$$\bar{M}_{\text{in-use}} = \sum_i \sum_j \sum_k \sum_l k v_{st}(i, j, k, l) \qquad \bar{M}_{\text{junk}} = \sum_i \sum_j \sum_k \sum_l j v_{st}(i, j, k, l)$$

$$\bar{M}_{\text{total}} = \sum_i \sum_j \sum_k \sum_l (j + k) v_{st}(i, j, k, l) \tag{3}$$

CPU utilisation rates can be calculated as

$$\rho_{\text{service}} = \sum_i \sum_{j \geq 1} \sum_k v_{st}(i, j, k, \text{service}) + \sum_i \sum_k v_{st}(i, 0, k, \text{service}) \min(1, r_i/s)$$

$$\rho_{\text{GC}} = \sum_i \sum_j \sum_k v_{st}(i, j, k, \text{GC}) \qquad\qquad \rho_{\text{total}} = \rho_{\text{service}} + \rho_{\text{GC}}$$

$$\tag{4}$$

In order to calculate the mean time of garbage collection intervals, we first need to calculate the average in-use memory level at the beginning of a garbage collection period.

$$\bar{M}_{\text{in-use at GC start}} = \sum_i \sum_k k v_{st}(i, N - k, k, \text{GC})/W_{\text{GC start}}, \text{ where} \tag{5}$$

$$W_{\text{GC start}} = \sum_i \sum_k v_{st}(i, N - k, k, \text{GC}); \tag{6}$$

then the mean time of garbage collection intervals is simply calculated as

$$\bar{T}_{GC} = \bar{M}_{\text{in-use at GC start}}/g \tag{7}$$

and the mean time of an entire cycle of service plus garbage collection can be calculated as

$$\bar{T}_{\text{total period}} = \bar{T}_{\text{GC period}}/\rho_{GC}. \tag{8}$$

For mean loss rate, we use the formula

$$\bar{L} = \sum_i \max((r_i - s), 0) v_{st}(i, N - 1, 0, \text{service}) +$$
$$\sum_i \sum_{j < N-1} v_{st}(i, j, N - j, GC) \max((r_i - g), 0) + \tag{9}$$
$$\sum_i \max((r_i - g), 0) v_{st}(i, N - 1, 1, GC),$$

and the mean loss ratio is

$$\bar{l} = \bar{L}/\bar{r}, \tag{10}$$

where $\bar{r}$ is the average rate of arrival.
The *effective rate of service* is

$$s_e = \frac{gs}{g + s}. \tag{11}$$

since each arrival needs to be served with rate s and (after some time) flushed with rate g.

Analysis of the average response time requires a more involved calculation.

3.1   Analysis of Average Response Time

For analysis of average response time, we assume the system is FIFO. Average response time is the total time spent in the system (spent with either service or waiting for service). It will also be referred to as delay.

The main idea is the following: when a tagged unit of data ("job") arrives during state (i, j, k, l), it will enter the queue. We consider this job as in position j within the queue, where each position corresponds to a unit segment within the queue. As the jobs are served, the tagged job will move ahead in the queue, eventually reaching position 1 and then being served.

The position of the tagged job within the queue as the system progresses is not included in the state of the system in the previously defined Markov chain. Instead, we represent it as the level in a quasi birth-death process (QBD) (see [14]), where the states are $((i, j, k, l), m)$, with m denoting the position of the tagged job within the queue (the level). (We remark that matrix-geometric

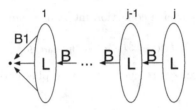

Fig. 3. QBD representation

methods are also a possible alternative to the QBD approach presented, see [20].)
Initially, a job arrives in state $((i, j, k, l), j)$ with probability

$$\pi(i, j, k, l) = \frac{v_{st}(i, j, k, l) r_i}{\sum_{i,j,k,l} v_{st}(i, j, k, l) r_i} \tag{12}$$

since in state (i, j, k, l) jobs arrive with rate r_i. π is understood as a row vector
of size $|\Omega|$.

All transitions of the original generator are partitioned into matrices B and
L (of size $|\Omega| \times |\Omega|$), with B corresponding to transitions that decrease the level,
that is, the transitions corresponding to service. L corresponds to the rest of the
transitions. In order to avoid listing all transitions again, we refer to Sect. 2.2;
from among all transitions listed there, all the transitions where *the third coordinate increases* go to B, while the rest of the transitions go to L (including
the negative values in the diagonal). The corresponding QBD represents the
progress of the tagged job along with the state of the entire system. See Fig. 3.
The process does not contain actual 'births', since the level may only decrease. In
such a system, let T_1 denote the (random) time it takes to go down one level to
some state (i', j', k', l'), assuming we started from state (i, j, k, l), and H denotes
its Laplace-transform:

$$H_{T_1}(s)_{(i,j,k,l),(i',j',k',l')} =$$
$$E(e^{-T_1 s} 1(\text{first backwards transition is to state } (i', j', k', l')) \tag{13}$$
$$|\text{starting from state } (i, j, k, l))$$

H can be calculated as follows [14]:

$$H_{T_1}(s) = (sI - L)^{-1} B.$$

The total delay T of the tagged job is equal to the time it takes to cross j
levels from initial distribution π and regardless of the end state (see also Fig. 3),
which has Laplace transform

$$H_T(s) = \sum_{i,j,k,l} \pi_{(i,j,k,l)} \cdot H_{T_1}^j(s) \cdot 1 \tag{14}$$

where $\mathbb{1}$ denotes the constant 1 column vector of size $|\Omega|$. Thus

$$E(T) = -\left.\frac{d}{ds}H_T(s)\right|_{s=0} = -\left.\frac{d}{ds}\sum_{i,j,k,l}\pi_{(i,j,k,l)}\cdot H_{T_1}^j(s)\right|_{s=0}\cdot\mathbb{1} =$$

$$-\left.\frac{d}{ds}\sum_{i,j,k,l}\pi_{(i,j,k,l)}\cdot((sI-L)^{-1}B)^j\right|_{s=0}\cdot\mathbb{1} =$$

$$-\sum_{i,j,k,l}\pi_{(i,j,k,l)}\cdot\left.\sum_{m=0}^{j-1}((sI-L)^{-1}B)^m(sI-L)^{-2}B((sI-L)^{-1}B)^{j-1-m}\right|_{s=0}\cdot\mathbb{1} =$$

$$-\sum_{i,j,k,l}\pi_{(i,j,k,l)}\cdot\sum_{m=0}^{j-1}((-L)^{-1}B)^m(-L)^{-2}B((-L)^{-1}B)^{j-1-m}\cdot\mathbb{1}. \tag{15}$$

Similarly,

$$E(T^2) = \left.\frac{d^2}{ds^2}H_T(s)\right|_{s=0} = \sum_{i,j,k,l}\left[\pi_{(i,j,k,l)}\cdot\sum_{m=1}^{j-1}\sum_{l=0}^{m-1}((-L)^{-1}B)^l(-L)^{-2}\times\right.$$

$$\times B((-L)^{-1}B)^{m-1-l}(-L)^{-2}B((-L)^{-1}B)^{j-1-m}+$$

$$\sum_{m=0}^{j-1}((-L)^{-1}B)^m(-2)(-L)^{-3}B((-L)^{-1}B)^{j-1-m}+$$

$$\left.\sum_{m=0}^{j-2}((-L)^{-1}B)^m(-L)^{-2}B\sum_{l=0}^{j-2-m}((-L)^{-1}B)^l(-L)^{-2}B((-L)^{-1}B)^{j-2-m-l}\right]\cdot\mathbb{1}. \tag{16}$$

(15) and (16) are explicit for $E(T)$ and $E(T^2)$, thus the mean and variance of the delay can be calculated. However, L and B are sparse matrices of size $|\Omega|\times|\Omega|$, which is typically large, so the actual calculations need special care. In the rest of this section, we sketch an efficient algorithm for the calculation of the formulas (15) and (16) for large Ω.

The first main point is that for large Ω, we only make calculations with vectors. To calculate (15), we start with the rightmost vector $\mathbb{1}$. Then, apart from summations, only 2 steps are repeated: either multiplication by B, which is feasible, or multiplication by $(-L)^{-1}$. The calculation of $(-L)^{-1}$ is infeasible, so to calculate $(-L)^{-1}v$ for some v, we solve $(-L)x = v$ instead. L has a special structure; we show that with a proper reordering of the states, L it will be upper block diagonal (with small block sizes), which allows $(-L)x = v$ to be solved block by block.

The ordering is as follows:

- The states for the same values of j, k, l will form blocks of size $|Q|$. The order within the block is irrelevant.
- For each value of j, k, the block for $l = 1$ comes before the block for $l = 2$.
- Then, for each value of j, the blocks are ordered in an increasing manner according to k (without changing the order of the blocks belonging to the same value of k).

– Then the blocks are ordered in a decreasing manner according to the value of j (without changing the order of the blocks belonging to the same value of j).

According to Fig. 1, this means that the last block is the bottom right corner of the GC triangle, preceded by the bottom right corner of the service triangle. Then the bottom rows of the two triangle follow from right to left, with blocks from the GC triangle and blocks from the service triangle alternating. Then the left of the rows follow from bottom to top.

The first two blocks (corresponding to the bottom right corners of each triangle) are special in the sense that L contains transitions between them in both directions. However, from all other blocks, L only contains transitions that go to later blocks (according to the above ordering), so in the above ordering, L is indeed block-upper-triangular, with a single diagonal block of size $2|Q|$ and all other diagonal blocks of size $|Q|$.

This allows us to solve $(-L)x = v$ for any v efficiently.

Starting from the vector $\mathbb{1}$, we keep multiplying by B and $(-L)^{-1}$ until we obtain the vectors $((-L)^{-1}B)^m(-L)^{-2}B((-L)^{-1}B)^{j-1-m} \cdot \mathbb{1}$ (from (15)). This process can be sped up by storing the vectors $((-L)^{-1}B)^j \cdot \mathbb{1}$ for separate values of j. Then the final summation can be made more efficient by pre-splitting π into vectors $\pi = \sum_j \pi_j$, where π_j only contains the elements of π whose second coordinate is j (that is, π_j corresponds to a single row in the service and GC triangles). Then

$$E(T) = \sum_j \pi_j \cdot \sum_{m=0}^{j-1} ((-L)^{-1}B)^m(-L)^{-2}B((-L)^{-1}B)^{j-1-m} \cdot \mathbb{1}. \tag{17}$$

(16) can be calculated efficiently using similar techniques (albeit with more steps). We do not go into further details due to lack of space here.

4   Experimental Results

4.1   Calculating Network Performance KPIs

The basis of the experimentation is a storm-based data processing system that uses reports from a large number of network elements (e.g. base stations) to calculate higher level network performance KPIs (key performance indicators). The topology of the processing system is a four-stage pipeline. We examine the first stage, called Parser, which parses the reports and retrieves the measurements from them.

The experimentation took place in the lab environment of Nokia, Bell Labs, with status reports stored in an HDFS storage and played back with real traffic timing. The processing software is implemented within the Apache Storm framework [2]. For monitoring, the Ganglia monitoring system was used [1]. Measurements were registered at intervals of length 500 ms.

4.2   Application of the Model

We apply the model to the Parser unit. First, the input data stream was approximated by a stationary Markov-modulated fluid model using k-means clustering to obtain the background Markov process with generator Q. Technically, input is given in discrete units (files), but the file size is relatively small compared to the total memory size.

Initial measurements showed that file size of the input data is proportional to both the amount of memory used during service, and also to the service time necessary. The corresponding constant factors were measured and are used as an input to the model. Service rate was also measured.

First, we are interested in the effect of discretisation: we model the same input process with several different setups of (N_U, N_V) pairs. The input parameters are (r and Q are not included in their entirety; input was clustered to 6 clusters):

$$s = 14.6\,\text{MB/s}, \ g = 64440\,\text{MB/s}, \ C = 40.81, M = 252\,\text{MB} \tag{18}$$
$$\bar{r} = 0.78\,\text{MB/s}, \ \max(r) = 20.9\,\text{MB/s}.$$

Service rate was measured using an artificially overloaded system, while the constant C was obtained by comparing the junk memory and the size of the incoming data. We note that the parameters in (18) reflect a relatively low load of the system. With a high load, certain processes such as memory swapping may be initiated which are not included in the model.

Table 1 contains the values of several performance measures obtained from the stationary analysis of the model for various (N_U, N_V) pairs.

Table 1. Effect of discretisation

	$(4, 8)$	$(10, 20)$	$(20, 40)$	$(10, 50)$
Mean period length	7.8282	7.8196	7.8193	7.8195
Utilisation	0.05421	0.05429	0.05429	0.05433
Mean loss ratio	1.17e−6	6.51e−10	1.69e−12	2.87e−13

The mean period length and the utilisation change very little as (N_U, N_V) are increased. On the other hand, the mean loss ratio is small and decreases rapidly as (N_U, N_V) increases. For the above input, it should be considered 0.

Analysis showed that the effect of the discretisation is relatively small, in other words, the model performs well with moderately large values of N_U and N_V (at least for utilisation and mean period length); from now on, we set $(N_U, N_V) = (10, 20)$ but with various inputs for actual validation.

From among the performance measures calculated in Sect. 3, we use the mean period length for validation with real life data. Mean period length is the mean time of an entire cycle of a service plus garbage collection period. The mean period length is easy to measure reliably: the real life monitoring system keeps count of the number of garbage collections over a sustained period of time. Several other performance measures are difficult to measure reliably: CPU usage relates to utilisation but may be distorted by other system processes. Loss ratio is known to be 0 from the actual monitoring, and this is approximated fairly well by the model, but relative error does not make sense in this case. Delay and the length of the queue was not possible to measure with the monitoring system.

The memory cap is slightly different for each run, ranging between 190 MB and 252 MB. The input process also varies slightly, with the minimal input rate 0, maximal input rate changing between 17 MB/s and 21 MB/s, and average input rate changing between 0.72 MB/s and 0.78 MB/s (Table 2).

Table 2. Validation of mean period length

Input run	1	2	3	4	5	6	7
From model	7.8196	8.2805	7.6990	7.2829	6.8598	6.6874	6.7157
Monitored	7.6585	8.1437	7.5303	7.1277	6.7030	6.5090	6.5090
Relative error	1.02%	1.02%	1.02%	1.02%	1.02%	1.03%	1.03%

Overall, the relative error is around 1%, with the model consistently overestimating the mean period length according to actual monitored results. The exact explanation and correction to the model is subject to further research, along with a more direct validation of the model. We also believe that the model presented models garbage collection on a realization level (not just stationary behaviour), but again, this is difficult to validate due to the fact that measurements made too often will distort the results themselves.

Close results in the literature are due to [4], but differences in the models (for example, description of the arrival process) make a direct comparison difficult.

5   Conclusion

The model is only applicable for a certain region of parameters. Under certain conditions, processes like memory swapping may be initiated. These are not included in the model.

The current model only includes one "type" of memory. However, in many memory management applications, there are "young" and "old" sections of the memory to store data for short and long term calculations. Such sections may be integrated in the model naturally with the expansion of the state space. This is subject to future work.

The computation of the stationary distribution (and the derived performance measures of Sect. 3) may be infeasible for very large values of N_U and N_V. Possible future work includes the application of dimension-reduction techniques based on tensor decomposition [17].

The model is sophisticated enough to allow modelling of a process on a realization level. This may be explored further.

Another natural option is to examine a transient version of the model; this would allow the examination of unstable systems as well.

An explicit solution for the original fluid model of Sect. 2.1 is also an interesting challenge.

Acknowledgment. We would like to thank Miklós Telek and Gábor Horváth for their valuable help and insight. This research is partially supported by the OTKA K123914 project.

References

1. Ganglia monitoring system. http://ganglia.sourceforge.net/. Accessed 08 May 2017
2. Apache Storm. http://storm.apache.org/. Accessed 08 May 2017
3. Bacon, D.F., Cheng, P., Rajan, V.T.: The metronome: a simpler approach to garbage collection in real-time systems. In: Meersman, R., Tari, Z. (eds.) OTM 2003. LNCS, vol. 2889, pp. 466–478. Springer, Heidelberg (2003). doi:10.1007/978-3-540-39962-9_52
4. Balsamo, S., Dei Rossi, G.-L., Marin, A.: Optimisation of virtual machine garbage collection policies. In: Al-Begain, K., Balsamo, S., Fiems, D., Marin, A. (eds.) ASMTA 2011. LNCS, vol. 6751, pp. 70–84. Springer, Heidelberg (2011). doi:10.1007/978-3-642-21713-5_6
5. Blackburn, S.M., Cheng, P., McKinley, K.S.: Oil and water? High performance garbage collection in Java with MMTk. In: Proceedings of the 26th International Conference on Software Engineering. IEEE Computer Society (2004)
6. Bolch, G., et al.: Queueing Networks and Markov Chains: Modeling and Performance Evaluation with Computer Science Applications. Wiley, Hoboken (2006)
7. Brodie-Tyrrell, W.: Surf: an abstract model of distributed garbage collection. Dissertation (2008)
8. Bux, W., Iliadis, I.: Performance of greedy garbage collection in flash-based solid-state drives. Perform. Eval. **67**(11), 1172–1186 (2010)
9. Detlefs, D., et al.: Garbage-first garbage collection. In: Proceedings of the 4th International Symposium on Memory Management. ACM (2004)
10. Gribaudo, M., Telek, M.: Fluid models in performance analysis. In: Bernardo, M., Hillston, J. (eds.) SFM 2007. LNCS, vol. 4486, pp. 271–317. Springer, Heidelberg (2007). doi:10.1007/978-3-540-72522-0_7
11. Jones, R., Lins, R.D.: Garbage Collection: Algorithms for Automatic Dynamic Memory Management. Wiley, New York (1996)
12. Jones, G.L., et al.: Fluid queue models of battery life. In: 2011 IEEE 19th Annual International Symposium on Modelling, Analysis, and Simulation of Computer and Telecommunication Systems. IEEE (2011)
13. Kwon, O., et al.: FeGC: an efficient garbage collection scheme for flash memory based storage systems. J. Syst. Softw. **84**(9), 1507–1523 (2011)
14. Latouche, G., Ramaswami, V.: Introduction to Matrix Analytic Methods in Stochastic Modeling. ASA-SIAM, Philadelphia (1999)
15. Li, Y., Lee, P.P.C., Lui, J.C.S.: Stochastic modeling and optimization of garbage collection algorithms in solid-state drive systems. Queueing Syst. **77**(2), 115–148 (2014)

16. Lowry, M.C.: A new approach to the train algorithm for distributed garbage collection. Dissertation (2004)
17. Kressner, D., Macedo, F.: Low-rank tensor methods for communicating Markov processes. In: Norman, G., Sanders, W. (eds.) QEST 2014. LNCS, vol. 8657, pp. 25–40. Springer, Cham (2014). doi:10.1007/978-3-319-10696-0_4
18. Medhi, J.: Stochastic Models in Queueing Theory. Academic Press, Cambridge (2002)
19. Nazarathy, Y., Weiss, G.: A fluid approach to job shop scheduling: theory, software and experimentation. J. Sched. **13**, 509–529 (2009)
20. Neuts, M.: Matrix-Geometric Solutions in Stochastic Models. An Algoritheoremic Approach. The Johns Hopkins University Press, Baltimore (1981)
21. Norcross, S.J.: Deriving distributed garbage collectors from distributed termination algorithms. Dissertation, University of St Andrews (2004)
22. Plainfossé, D., Shapiro, M.: A survey of distributed garbage collection techniques. In: Baler, H.G. (ed.) IWMM 1995. LNCS, vol. 986, pp. 211–249. Springer, Heidelberg (1995). doi:10.1007/3-540-60368-9_26
23. Schoeberl, M.: Real-time garbage collection for Java. In: Ninth IEEE International Symposium on Object and Component-Oriented Real-Time Distributed Computing (ISORC 2006). IEEE (2006)
24. Lakatos, L., Szeidl, L., Telek, M.: Introduction to Queueing Systems with Telecommunication Applications. Springer, New York (2013)
25. Van Houdt, B.: A mean field model for a class of garbage collection algorithms in flash-based solid state drives. In: ACM SIGMETRICS Performance Evaluation Review. vol. 41, no. 1. ACM (2013)
26. Van Houdt, B.: Performance of garbage collection algorithms for flash-based solid state drives with hot/cold data. Perform. Eval. **70**(10), 692–703 (2013)
27. Wilson, P.R.: Uniprocessor garbage collection techniques. In: Bekkers, Y., Cohen, J. (eds.) IWMM 1992. LNCS, vol. 637, pp. 1–42. Springer, Heidelberg (1992). doi:10.1007/BFb0017182
28. Yang, Y., Zhu, J.: Analytical modeling of garbage collection algorithms in hotness-aware flash-based solid state drives. In: 2014 30th Symposium on Mass Storage Systems and Technologies (MSST). IEEE (2014)

A Simple Approximation for the Response Times in the Two-Class Weighted Fair Queueing System

Dhari Ali Mahmood[1,2,3] and Gábor Horváth[1,2(✉)]

[1] Department of Networked Systems and Services,
Budapest University of Technology and Economics, Budapest, Hungary
[2] MTA-BME Information Systems Research Group,
Magyar Tudósok Krt. 2, Budapest 1117, Hungary
{dhariali,ghorvath}@hit.bme.hu
[3] University of Technology, Baghdad, Iraq

Abstract. The weighted fair queueing (WFQ) service discipline provides a flexible way to share bandwidth among two or more traffic classes. Some variants of the basic WFQ principle are used in the practice in computer networks in routers, switches, etc. Unfortunately, the analytical modeling of the related queues turned out to be notoriously difficult. This paper presents approximation expressions for the mean response times in a two-class (ideal) WFQ system with Poisson arrival process and exponentially distributed service times. The approximation is based on simulation. The results are very simple, explicit, yet reasonably accurate, ideal to use in self organizing networks where the weights associated with the different traffic classes need to be recalculated to adapt to the changing network conditions.

1 Introduction

In computer and telecommunication networks the overall traffic is a mixture of packet flows having different quality demands. Some packets are urgent, while some others can tolerate delay better. Most modern communication protocols have a field in the packet header indicating to which class the packet is belonging to (like the class of service (CoS) field in the Ethernet frame header and the DiffServ code point (DSCP) in the IP header). Packet schedulers in the network devices (switches, routers) need to take this information into account to provide the necessary quality of service.

A popular multi-class scheduling discipline for this purpose is the weighted fair queueing (WFQ) service. In such systems the packets belonging to different traffic classes are stored in separate queues before they get transmitted. The total service capacity is shared among the classes according to the weights associated with the queues: The higher the weight of a traffic class is, the higher service rate it gets. The WFQ schedulers are work conserving, thus the total service capacity is always distributed among the classes that are currently active. The weights provide a flexible way to express the importance of the traffic classes.

© Springer International Publishing AG 2017
N. Thomas and M. Forshaw (Eds.): ASMTA 2017, LNCS 10378, pp. 125–137, 2017.
DOI: 10.1007/978-3-319-61428-1_9

The fluid-based version of the scheduler, where the customers are infinitesimally small (considered as fluid drops), is often called generalized processor sharing (GPS), while the variant with discrete customers, also studied in the paper, is called weighted fair queueing. According to the ideal weighted fair queueing (also referred to as the coupled processor model in [3,8,12]) the multiple traffic classes can be served simultaneously, at the reduced service rate associated to them. This ideal WFQ is, however, impossible to implement in a real situation. Several packet-based approximations of the ideal WFQ appeared in the practice, including the Virtual Clock [14], Self-Clocked Fair Queueing [4], Deficit Round Robin [10], etc.

Although these WFQ-like schedulers are very popular, there are very few analytical results available in the literature. In the simplest scenario with Poisson arrival process and exponentially distributed service times, the Markov chain representing the number of customers in the system (class-wise) has a simple, regular structure, still, its stationary solution turned out to be a notoriously difficult problem. The only exact result we are aware of is [5], where the generating function was derived (the mathematical apparatus used in that paper demonstrates how difficult the problem is). Several approximations appeared as well to provide simpler, more tractable solution of the WFQ system. The result in [6] is based on the decoupling of the queues, while the QBD structure of the Markov chain is exploited in [1]. The idea in [9] is to transform the WFQ system to a priority queue.

In this paper our aim is to provide a very simple, explicit approximation for the two-class ideal WFQ system, based on simulation results and curve fitting. Similar approach has been followed many times in the past: the KLB formula [7] for the approximation of the waiting time in $G/G/1$ queues and the formulas in [13] to approximate various properties of the departure process of $G/G/1$ queues were both successful and widely used results. Our approach is somewhat similar to [9], but that paper considers a slightly different system where the service of packets can not be preempted, and a step of the procedure needs the numerical solution of an equation. Our formulas are explicit, contain only basic operations, and can be easily implemented in a network device, making it possible to recalculate the weights of the classes if the traffic situation changes.

2 Model Definition

In this paper we consider the two-class weighted fair queueing system. The customers are arriving according to a Poisson process with parameters λ_1 and λ_2, and are directed to two separate queues according to their class. The service times are exponentially distributed with (class independent) parameter μ. The server is shared among the two customer classes, controlled by weights w_1 and w_2. According to the ideal weighted fair queueing policy considered in the paper, both the class 1 and class 2 queues are served in parallel, if both kinds of customers are present in the system: class 1 is served with rate $\mu \cdot w_1/(w_1 + w_2)$, while class 2 is served with rate $\mu \cdot w_2/(w_1 + w_2)$. If one of the queues is idle then the total service capacity is given to the other class.

The amount of work brought by class k customers to the system is $\rho_k = \lambda_k/\mu$. In this paper we assume that the system is stable, hence for the total utilization $\rho = \rho_1 + \rho_2$ we have that $\rho < 1$.

The asymmetry of the utilization of the two customer classes can be characterized many ways. We found that the measure

$$r = \frac{\rho_1 - \rho_2}{\rho}, \tag{1}$$

$r \in (-1, 1)$, turned out to be a good choice, making the forthcoming expressions simpler (Table 1).

Table 1. Notation and parameters used.

Parameter	Definition
μ	Service rate
λ_1, λ_2	Arrival rate of class 1 and class 2 customers
w_1, w_2	Weight of class 1 and class 2 customers
ρ_1, ρ_2	Utilization of class 1 and class 2 customers
ρ	The total utilization of the system
r	The asymmetry of the utilization
$E(T_1), E(T_2)$	Response time (waiting+service time) of class 1 and class 2 customers

2.1 Analytical Results Used in the Paper

To approximate the mean response time in the WFQ system, we are going to utilize the results of two closely related two-class queueing systems, that have exact mean response time results available.

One of these systems is the two-class FCFS queue. In this system there is no capacity sharing and all demands are served according to the global arrival order independent on the class. The mean response time is given by [2]

$$E(T_{FCFS}) = \frac{1}{\mu - \lambda_1 - \lambda_2}. \tag{2}$$

The second queueing system necessary to our approximation is the two-class preemptive priority queue. Observe that the ideal WFQ server investigated in this paper exhibits a kind of preemptive behavior: when a customer arrives to an idle queue, the service rate of the other class gets reduced immediately. When one of the weights, w_1 or w_2 is zero, then the WFQ behaves like a preemptive (resume) priority queue. If $w_1 = 0$, class 1 plays the role of the low priority class with mean response time given by

$$E(T_{Prio}) = E(T_{FCFS})\frac{1}{1 - \rho_2} \tag{3}$$

Finally, the conservation law [2]

$$\rho_1 E(T_1) + \rho_2 E(T_2) = \rho E(T_{FCFS}) \tag{4}$$

allows us to focus on one of the classes only, the mean response time for the opposing class can be calculated from (4).

2.2 Simulation of the System

Due to the naming confusion and the existence of many variants of the WFQ system, and since we rely on simulation results heavily in the paper, we briefly discuss the simulation of the queue studied in this paper. Algorithm 1 provides the simplified simulation algorithm in a discrete event simulation system[1].

Algorithm 1. Discrete event simulation of the WFQ system

Event $end\_of\_service_i$:
    collect (current time - arrival time) to response time statistics
    remove customer from $queue_i$
    if $queue_i$ is empty **then**
        call RESCHEDULESERVICETIMES
    else
        select next customer in $queue_i$
        service time $\leftarrow$ $\text{Exp}(\mu \cdot share_i)$
        schedule $end\_of\_service_i$ **to** current time + service time
    end if
End
Event $arrival\_of\_class_i$:
    if $queue_i$ is empty **then**
        add new customer to $queue_i$
        call RESCHEDULESERVICETIMES
    else
        add new customer to $queue_i$
    end if
End
procedure RESCHEDULESERVICETIMES
    for every non-empty $queue_i$ **do**
        $share_i \leftarrow w_i / \sum\limits_{\substack{\forall j: queue_j \\ \text{not empty}}} w_j$
        cancel event $end\_of\_service_i$
        service time $\leftarrow$ $\text{Exp}(\mu \cdot share_i)$
        schedule $end\_of\_service_i$ **to** current time + service time
    end for
end procedure

In this algorithm there are two kinds of events to handle: arrival and service events. There are as many service events scheduled at the same time as many

---

[1] Our implementation is based on OmNet++ [11].

busy queues there are in the system. These events need to be re-scheduled when the busy state of the queues changes, that can occur in two situations: when a customer arrives into an empty queue and when a customer leaves its queue empty. In case of exponentially distributed service times re-scheduling the service events is simple, the memory-less property can be exploited.

3 The Analysis of the Response Times

3.1 The Concept of the Approximation

Due to the conservation law (4) it is enough to focus on a single customer class, class 1, the mean response time for the other traffic class can be expressed from (4). An other feature of the system that we are going to exploit is that the two weight parameters w_1, w_2 defining the system are redundant. In the sequel, we are going to set $w_2 = 1$ and investigate the behavior of the system as the function of w_1.

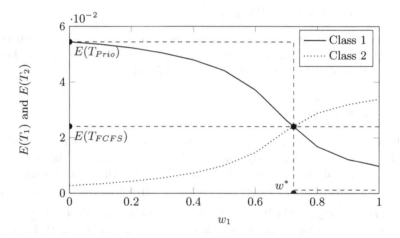

Fig. 1. The mean response times as the function of w_1 ($\mu = 0.0012$)

Figure 1 depicts the mean response times as the function of w_1 in a particular example ($\rho_1 = 0.39, \rho_2 = 0.55$). Observe that if $w_1 = 0$ then the system behaves like a preemptive priority queue with class 1 being the low priority class, hence $E(T_1) = E(T_{Prio})$. At the other hand, when $w_1 \to \infty$, class 1 has exclusive access to the service capacity. The point where the curves of class 1 and class 2 meet plays an important role in our approximation. In this point $E(T_1) = E(T_2)$ holds, more precisely, (4) implies that $E(T_1) = E(T_2) = E(T_{FCFS})$. The weight belonging to this point is denoted by w^* in the sequel. Based on this point the plot of class 1 on the figure can be divided to two rectangular regions (denoted by dashed lines). Due to the symmetry of the system, we assume that $w_1 \leq w^*$

holds (the role of the two classes can be swapped in the opposite case), we are going to study only this case in the rest of the paper, hence our aim is to approximate the behavior in the rectangle on the left.

Our approximation for the response times consists of two components:

- *The approximation of w^*.* This is the only unknown parameter to fully characterize the region marked by dashed lines in Fig. 1. The top left point is given by $w_1 = 0, E(T_1) = E(T_{prio})$, and the bottom right point is located at $w_1 = w^*, E(T_1) = E(T_{FCFS})$.
- *The approximation of the shape of the response time curve.* Based on many simulation experiments we found that w^* is very close to the inflection point in most of the cases (except if the utilization is extremely low). Hence, $E(T_1)$ inside the dashed region is typically monotonous. The bend of the curve (referred to as the "shape parameter" in the sequel) depends on ρ_1 and ρ_2, and it is also subject to approximation.

The next two subsections present the approximation of these two parameters.

3.2   Approximating the Weight w^*

We have studied the behavior of w^* as the function of $\rho = \rho_1 + \rho_2$ with different r parameters (r characterizes the asymmetry, see (1)). We found that at the two extreme values of ρ the $w^*(\rho, r)$ tends to specific values:

- At $\rho \to 0$ w^* tends to 1,
- at $\rho \to 1$ w^* tends to $\rho_1/\rho_2 = \frac{1+r}{1-r}$.

The latter relation can be intuitively justified as follows[2]: When ρ is almost one, the system is continuously busy so the class-1 queue is like an $M/M/1$ queue with arrival rate λ_1 and service rate $\frac{\mu w_1}{w_1+1}$. So, the mean response time of class-1 customers is $E(T_1(w_1)) = \frac{w_1+1}{w_1(\mu-\lambda_1)-\lambda_1}$. To obtain w^* one has to solve $E(T_1(w_1)) = \frac{1}{\mu-\lambda_1-\lambda_2}$, which, after a few calculations leads to $w_1 = \frac{1-\rho_2}{\rho_2} = \frac{\rho_1}{\rho_2}$.

To make the visual comparison easier, we scale w^* to the $[0, 1]$ domain by introducing $\omega^*(\rho, r)$ as

$$\omega^*(\rho, r) = \frac{w^*(\rho, r) - \frac{\rho_1}{\rho_2}}{1 - \frac{\rho_1}{\rho_2}} = \frac{w^*(\rho, r) - \frac{1+r}{1-r}}{1 - \frac{1+r}{1-r}}. \tag{5}$$

Figure 2 depicts the shape of $\omega^*(\rho, r)$ as the function of ρ at various settings of r. According to our simulation experiments, these curves are (almost) symmetric to the $f(x) = x$ line. We found a family of functions, having the same symmetry, suitable for the approximation: the $f(x) = \frac{x-1}{cx-1}$ function. The c parameter of this function controls the shape: how much the curve bends towards the upper right corner of the rectangle defined by corners $(0, 0)$ and $(1, 1)$. Shape parameter c

---

[2] We would like to thank the anonymous review for the intuitive explanation presented.

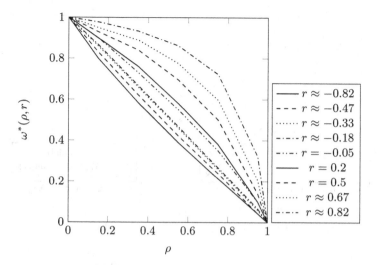

Fig. 2. The parameter $\omega^*(\rho, r)$ at various ρ and r values

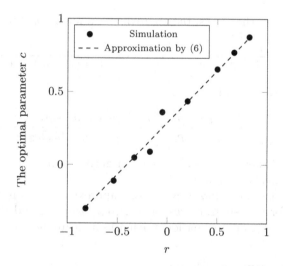

Fig. 3. The optimal shape parameter c for the approximation of $\omega^*(\rho, r)$

can be negative as well, in this case the curve bends towards the lower left corner instead.

For each r setting we determined the optimal shape c value leading to the most accurate approximation. Plotting these parameters as the function of r leads to a nearly linear function (Fig. 3)

$$c = 0.71r + 0.29. \tag{6}$$

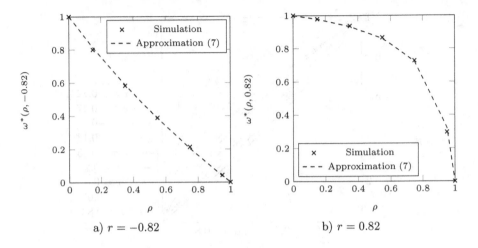

Fig. 4. The accuracy of the approximation of parameter ω^*

Putting together the pieces, the approximation for $\omega^*(\rho, r)$ is

$$\omega^*(\rho, r) = \frac{\rho - 1}{c\rho - 1} = \frac{\rho - 1}{(0.71r + 0.29)\rho - 1} = \frac{\rho - 1}{0.71(\rho_1 - \rho_2) + 0.29\rho - 1}, \quad (7)$$

finally, the approximation for w^* is

$$w^* = \omega^*(\rho, r)(1 - \frac{\rho_1}{\rho_2}) + \frac{\rho_1}{\rho_2} = \left(1 - \frac{\rho_1}{\rho_2}\right)\frac{\rho_1 + \rho_2 - 1}{\rho_1 - 0.42\rho_2 - 1} + \frac{\rho_1}{\rho_2}. \quad (8)$$

Note that $\omega^*(\rho, r)$ (hence w^*) is always non-negative in the stability region since the minimum value is given at $\rho = 1$.

Figure 4 demonstrates how accurate this approximation is in two cases, for $r = -0.82$ and for $r = +0.82$, corresponding to $\rho_1/\rho_2 = 0.1$ and 10, respectively (the dashed line is the approximation, the marks indicate the simulation results).

3.3 Approximating the Shape

Having w^* approximated, the next element of the solution is to approximate $E(T_1)$ when $w_1 \in [0, w^*]$ (the curve in the upper left dashed rectangle in Fig. 1). In this region $E(T_1)$ takes values between $E(T_{Prio})$ and $E(T_{FCFS})$. Let us scale this region to the $[0, 1]$ domain by defining $\hat{E}(T_1)$ as

$$\hat{E}(T_1) = \frac{E(T_1) - E(T_{FCFS})}{E(T_{Prio}) - E(T_{FCFS})} \quad (9)$$

and investigate its dependence on the parameters of the system. Depending on the class 1 and class 2 load the curve representing $\hat{E}(T_1)$ as the function of w_1/w^* bends towards the lower left or towards the upper right corner of the unit rectangle. The function $f(x) = \frac{x-1}{c\,x-1}$, introduced and used in the previous section, turned out to be suitable to approximate this curve as well. As before, the question is how to set the shape parameter c to make the approximation accurate.

First we studied the symmetric case with $\rho_1 = \rho_2$ ($r = 0$). Investigating the plot depicting the optimal shape parameter as the function of the total load $\rho = \rho_1 + \rho_2$ we found that it changes between -1 and 1 and that it can be approximated by $c = 2\rho^{2/3} - 1$ very accurately (see Fig. 5).

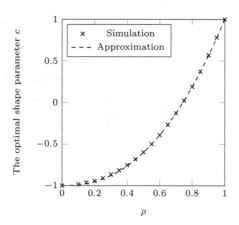

Fig. 5. The optimal shape parameter as the function of the total load

Hence, we were looking for the approximation in the non-symmetric ($\rho_1 \neq \rho_2$) case in the form of $c = 2\rho^{g(r)} - 1$ as well. The empirical analysis of the exponent revealed that

$$g(r) = 6 \cdot |r - 0.25|^{3.2} + 2 \tag{10}$$

is a relatively accurate approximation of the simulation results with less than 5% error (see Fig. 6), although it gives $g(0) = 2.071$ instead of $2/3$ for $r = 0$.

Altogether, the mean scaled response time $\hat{E}(T_1)$ as the function of the scaled weight w_1/w^* is approximated by

$$\hat{E}(T_1) = \frac{w_1/w^* - 1}{(2\rho^{6 \cdot |r - 0.25|^{3.2} + 2} - 1)w_1/w^* - 1}. \tag{11}$$

The complete algorithm including the selection of the role of class 1 and the approximation of both mean response times is presented in Algorithm 2. While the resulting formula is explicit, the algorithm brakes down the solution to multiple steps for simplicity.

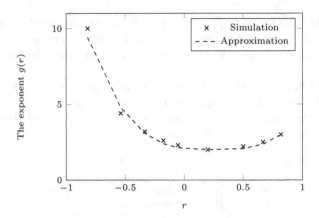

Fig. 6. The approximation of the exponent $g(r)$

Algorithm 2. The approximation of the mean response time
---

function $E(T_1), E(T_2) = \text{WFQRESPONSETIME}(\lambda_1, \lambda_2, \mu)$

  $\rho_1 \leftarrow \lambda_1/\mu, \quad \rho_2 \leftarrow \lambda_2/\mu, \quad \rho \leftarrow \rho_1 + \rho_2$

  $w^* \leftarrow \left(1 - \frac{\rho_1}{\rho_2}\right) \frac{\rho_1 + \rho_2 - 1}{\rho_1 - 0.42\rho_2 - 1} + \frac{\rho_1}{\rho_2}$

  if $w_1/w_2 < w^*$ **then**

   $r \leftarrow (\rho_1 - \rho_2)/\rho$

   $c \leftarrow 2\rho^{6 \cdot |r - 0.25|^{3.2} + 2} - 1$

   $w \leftarrow w_1/w_2$

   $E(T_{FCFS}) \leftarrow \frac{1}{\mu_1 - \lambda_1 - \lambda_2}$

   $E(T_{Prio}) \leftarrow E(T_{FCFS})/(1 - \rho_2)$

   $\hat{E}(T_1) \leftarrow (w/w^* - 1)/(c \cdot w/w^* - 1)$

   $E(T_1) \leftarrow \hat{E}(T_1)E(T_{Prio}) + (1 - \hat{E}(T_1))E(T_{FCFS})$

   $E(T_2) \leftarrow (\rho E(T_{FCFS}) - \rho_1 E(T_1))/\rho_2$

  else

   $r \leftarrow (\rho_2 - \rho_1)/\rho$

   $c \leftarrow 2\rho^{6 \cdot |r - 0.25|^{3.2} + 2} - 1$

   $w^* \leftarrow 1/w^*$

   $w \leftarrow w_2/w_1$

   $E(T_{FCFS}) \leftarrow \frac{1}{\mu_1 - \lambda_1 - \lambda_2}$

   $E(T_{Prio}) \leftarrow E(T_{FCFS})/(1 - \rho_1)$

   $\hat{E}(T_2) \leftarrow (w/w^* - 1)/(c \cdot w/w^* - 1)$

   $E(T_2) \leftarrow \hat{E}(T_2)E(T_{Prio}) + (1 - \hat{E}(T_2))E(T_{FCFS})$

   $E(T_1) \leftarrow (\rho E(T_{FCFS}) - \rho_2 E(T_2))/\rho_1$

  end if

  return $E(T_1), E(T_2)$

end function

---

4 Numerical Results

In this section we demonstrate the behavior of our approximation method and compare it with the procedure published in [6]. This comparison is not completely

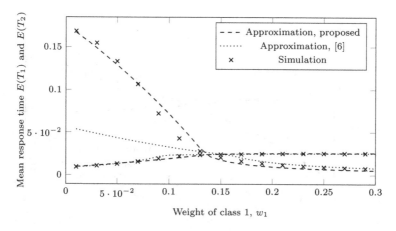

Fig. 7. The worst results obtained by the approximation, $\rho = 0.95, r = -0.82$ ($\mu = 0.0012$)

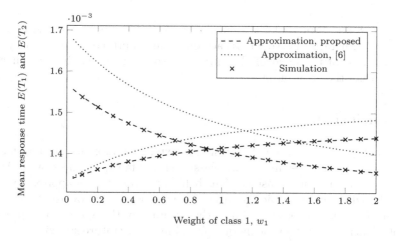

Fig. 8. Comparison with parameters $\rho = 0.15$ and $r = -0.33$ ($\mu = 0.0012$)

fair, though, since [6] considers a more general system where the inter-arrival and service times can be non-exponential as well.

In general, the proposed approximation managed to achieve very accurate results. In the extreme cases, when the utilization is high and the load is very asymmetric, the accuracy is worse, while in the more "balanced" cases the accuracy is better. Among the scenarios we investigated, the results were the worst with parameters $\rho = 0.95, r = -0.82$. The mean response times as the function of w_1 are depicted in Fig. 7. The reason of the sub-optimal performance is that under such a high load the inflection point of the curve does not coincide with w^*. However, the results are still much better than the ones obtained by [6].

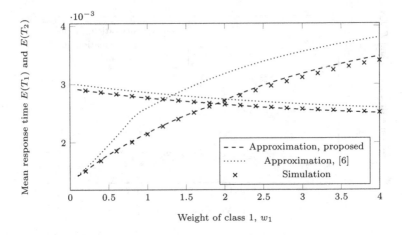

Fig. 9. Comparison with parameters $\rho = 0.55$ and $r = 0.67$ ($\mu = 0.0012$)

Figures 8 and 9 present the typical accuracy of the proposed method. The weights w^*, where the curves cross each other, are captured almost exactly. The approximation of the bend of the curve has some error, but it is much more accurate than the error of [6].

5   Conclusion

In this paper we have presented a simple explicit approximation formula for the mean response times in the two-class weighted fair queueing system. While there are some queueing considerations behind the results, the approximation is mostly based on an algebraic approach. Some decisions on how to approximate the behavior seem sometimes ad-hoc, the accuracy of the approximation is reasonable, much better than the method found in the literature studying the same system.

Acknowledgment. Dhari Ali Mahmood would like to thank to the Tempus Public Foundation (TPF) – Stipendium Hungaricum program and University of Technology – Iraq for the support for his PhD scholarship.

References

1. Al-Sawaai, A., Awan, I., Fretwell, R.: Analysis of the weighted fair queuing system with two classes of customers with finite buffer. In: International Conference on Advanced Information Networking and Applications Workshops, WAINA 2009, pp. 218–223. IEEE (2009)
2. Bolch, G., Greiner, S., de Meer, H., Trivedi, K.S.: Queueing Networks and Markov Chains: Modeling and Performance Evaluation with Computer Science Applications. Wiley, Hoboken (2006)

3. Fayolle, G., Iasnogorodski, R.: Two coupled processors: the reduction to a Riemann-Hilbert problem. Probab. Theory Relat. Fields **47**(3), 325–351 (1979)

4. Golestani, S.J.: A self-clocked fair queueing scheme for broadband applications. In: 13th Proceedings IEEE Networking for Global Communications, INFOCOM 1994, pp. 636–646. IEEE (1994)

5. Guillemin, F., Pinchon, D.: Analysis of generalized processor-sharing systems with two classes of customers and exponential services. J. Appl. Probab. **41**(03), 832–858 (2004)

6. Horváth, G., Telek, M.: An approximate analysis of two class WFQ systems. In: Workshop on Preformability Modeling of Computer and Communication Systems-PMCCS, pp. 43–46. Citeseer (2003)

7. Kraemer, W., Langenbach-Belz, M.: Approximate formulae for the delay in the queueing system GI/G/1. In: Proceedings of the 8th International Teletraffic Congress, pp. 235–1 (1976)

8. Resing, J.: A tandem queueing model with coupled processors. Oper. Res. Lett. **31**(5), 383–389 (2003)

9. Shortle, J.F., Fischer, M.J.: Approximation for a two-class weighted fair queueing discipline. Perform. Eval. **67**(10), 946–958 (2010)

10. Shreedhar, M., Varghese, G.: Efficient fair queuing using deficit round-robin. IEEE/ACM Trans. Netw. **4**(3), 375–385 (1996)

11. Varga,A., Hornig, R.: An overview of the omnet++ simulation environment. In: Proceedings of the 1st international conference on Simulation tools and techniques for communications, networks and systems & workshops, p. 60, ICST (Institute for Computer Sciences, Social-Informatics and Telecommunications Engineering) (2008)

12. Vitale, C., Rizzo, G., Rengarajan, B., Mancuso, V.: An analytical approach to performance analysis of coupled processor systems. In: 2015 27th International Teletraffic Congress (ITC 27), pp. 89–97. IEEE (2015)

13. Whitt, W.: The queueing network analyzer. Bell Labs Tech. J. **62**(9), 2779–2815 (1983)

14. Zhang, L.: Virtual clock: a new traffic control algorithm for packet switching networks. In: ACM SIGCOMM Computer Communication Review, vol. 20, pp. 19–29. ACM (1990)

Application of a Particular Class of Markov Chains in the Assessment of Semi-actuated Signalized Intersections

Francisco Macedo[1,3], Paula Milheiro-Oliveira[1,2(✉)], António Pacheco[3], and Maria Lurdes Simões[2,4]

[1] CMUP, Universidade do Porto,
Rua do Campo Alegre, 687, 4169-007 Porto, Portugal
francisco.quartin@gmail.com
[2] Faculdade de Engenharia, Universidade do Porto,
Rua Dr. Roberto Frias, s/n, 4200-465 Porto, Portugal
{poliv,lurdes.simoes}@fe.up.pt
[3] CEMAT and Instituto Superior Técnico, Universidade de Lisboa,
Av. Rovisco Pais, 1, 1049-001 Lisboa, Portugal
apacheco@math.tecnico.ulisboa.pt
[4] CONSTRUCT, Faculdade de Engenharia, Universidade do Porto,
Rua Dr. Roberto Frias, s/n, 4200-465 Porto, Portugal

Abstract. We investigate a queuing model for a signalized intersection regulated by semi-actuated control in a urban traffic network. Modelling the queue length and the delay of vehicles for this type of traffic, characterized by variable durations of the green signal, is crucial to evaluate the performance of traffic intersections. Additionally, determining the size of the extensions of the green signal is also relevant. The traffic systems addressed in the paper have the particularity that the server remains active (green signal) for a period of time that depends on the number of vehicles waiting at the intersection. This gives rise to an $M/D/1$ queuing system with a server that occasionally takes vacations (red signal), for which we compute the long-run mean delay of vehicles, mean queue length and mean duration of the green signal. We consider a case study and compare the results obtained from the proposed queueing model with those obtained by using a microsimulation model. The formulas derived for the performance measures are of interest for traffic engineers, since the existing alternative formulas are subject to strong criticism.

1 Introduction

The last decades of research on the theory of signalized traffic intersections put a lot of emphasis on estimation methods of delays and queue lengths at individual intersections regulated by actuated control and on the strategies that can be designed upon the results of such estimation and on the analysis of traffic characteristics. The performance of signalized intersections is indeed usually measured by the mean queue length and the mean delay (sojourn time in system) of vehicles.

© Springer International Publishing AG 2017
N. Thomas and M. Forshaw (Eds.): ASMTA 2017, LNCS 10378, pp. 138–151, 2017.
DOI: 10.1007/978-3-319-61428-1_10

Different approaches to the estimation problem can be found in the literature. The approach based on microscopic simulation models, essentially car-following models (see, e.g., [2,16,20]), presents some important disadvantages since, in spite of the fact that they mimic quite well the behaviour of traffic in real world, they need to be fed with a lot of parameters, not easily known or measured in practice, and require a considerable computational effort. Popular models like the HCM model [18] and Webster's model [25] are known to have also some drawbacks.

As an alternative, this paper explores the use of queueing theory in order to obtain the performance measures just mentioned above. The main difficulties involved in such an approach come from the need of a good characterization of the circulating vehicles and drivers, and from the fact that the cyclic deactivation of the server (the red signal) has to be incorporated in the behaviour of the queueing system. In the work published in [15] we have addressed pre-timed control intersections. However actuated or semi-actuated traffic signals are generally more efficient, since they better accommodate fluctuation of vehicle arrivals as they are able to adapt the green time given to a traffic stream according to demand, by incorporating the possibility of extending the green signal (see e.g. [23]).

The paper by Lin *et al.* [13] explores simple probabilistic arguments to obtain the mean duration of the green signal in semi-actuated controlled intersections, but their approach is restricted to small volumes of traffic in the secondary street, smaller than 500 vehicles per hour. Even in the case of Poisson vehicle arrivals, models like $M/D/1$ and $M/D^X/1$ do not correctly describe the deactivation of the server, taking place when the signal changes from green to red. In fact, queueing systems with server vacations (see [3] for a survey) are a more convenient way of modelling the stochastic behaviour of the traffic system (see also previous work in [7,8,24] for the case of pre-timed control).

Signalized traffic intersections have similarities with polling systems (see e.g. [21] or [22] for an overview on polling systems), where a single server is handling two queues and switches between them according to some control rule. In the case of semi-actuated signalized intersections, queues are attended by the server during given periods of time, which may have random duration – at least for one of the queues. However, as far as we know, the diversity of polling systems found in the literature do not encompass the specificity of the semi-actuated signalized traffic addressed in the paper. Several authors (see e.g. [5,6,11] or [1]) stress the fact that systems characterized by time limited service disciplines, as it is the case for semi-actuated signal intersections, should not be expected to have closed formulas for the expected customer waiting time. The papers just cited focus on cases of exponential or phase-type service times, which do not apply to signalized traffic. However, time limited server systems are often used when in presence of heavy loaded queues that tend to monopolize the server, leaving lightly loaded queues with a negligible part of the service time.

In this paper, we consider a semi-actuated isolated signalized intersection, meaning that the mechanism that triggers red times relies on the evolution of the traffic demand, leading to green times of random duration. Specifically, the green time is extended, from a fixed minimum duration, in case there are vehicles waiting at the intersection at the end of a minimum green time period. Additional individual extensions of the green time by T seconds are performed if the time interval between arriving vehicles remains smaller than T seconds, up to the green time reaching a maximum pre-fixed total duration. For implementing the green time extension mechanism, a sensor located a couple of meters before the stop line is responsible for the detection of vehicle at the intersection.

We model the semi-actuated signalized intersection as an $M/D/1$ queueing model with server vacations, in which clients (vehicles) are served in a first-in first-out (FIFO) regime. The server starts a vacation of fixed duration as soon as a red time initiates. As described in the previous paragraph, server working periods, corresponding to green times, have random duration. We explore in the paper the specific nature of the resulting $M/D/1$ server vacation queue, and in particular its Markov regenerative structure, to characterize the distributions of queue length, vehicle delay, and duration of the green signal in the long-run regime. Our approach is different from that of [12], which relies on the derivation of a functional equation for the system behavior and its solution by means of a numerical technique based on Laguerre-function approximations. We compare the results obtained for the derived long-run measures with those obtained by applying a microscopic simulation model (see [19]). Our main contribution lies in providing expressions for the means of waiting time of drivers, length of queue at the intersection, and total duration of the green signal, which are of interest for traffic engineers.

The paper is organized as follows. The assumptions made and the Markov chain model that is used in the paper for investigating semi-actuated signalized traffic intersections are introduced in Sect. 2. The main results on long-run performance measures for semi-actuated signalized traffic intersections are included in Sect. 3, and a case study that is used to validate the results obtained from the proposed model is presented in Sect. 4. The paper ends with some brief conclusions drawn in Sect. 5.

2   The Signalized Intersection Traffic Model

A signalized intersection regulated by semi-actuated control is assumed to be a traffic server system for which each vehicle arriving at the intersection during a green (light) period has to wait if there are vehicles in front of it, or if arriving during a red (light) period. In a detailed way, we consider a model for a signalized intersection having the following specifications, with time in seconds:

- Vehicles arrive at the intersection according to an homogeneous Poisson process with rate λ, and are served one by one in order of arrival.
- The intersection possesses infinite vehicle waiting capacity, and the light alternates between green and red periods.
- The service time of a vehicle is constant and equal to T, and services are initiated during green periods at instants that are integer multiples of T.

- Red periods have constant duration of value RT, and green periods have random durations, taking values on the set

$$\{MT, (M+1)T, \ldots, GT\}$$

such that: starting from an initial interval of duration MT for a green period, successive extensions of length T of the green period occur if there are vehicles to be served at the intersection at the end of the interval, with extensions being allowed only up to the point when the length of the green period reaches the corresponding maximum duration of GT.

Note that T is an arbitrary positive constant that denotes the time that a vehicle spends to move through the intersection, i.e., its service time, R and M are positive integers, and $G - M$ is a nonnegative integer number denoting the maximum number of extensions of T seconds that are allowed to be performed in green periods. Our assumptions imply that signal cycles have maximum duration $(G + R)T$, and are divided in a server working period of minimum length MT and maximum length GT, corresponding to a green period, followed by a server vacation period of fixed length RT, corresponding to a red period.

We should stress that the approach that will be followed in the paper could be adapted with small effort to accommodate: vehicles arriving at the intersection according to a non-homogeneous compound-Poisson process; the intersection having finite vehicle waiting capacity, and group service of vehicles – with a maximum size group being allowed, as considered in [8]. The time discretization, with time step T, which is implicit in the Markov chain that we will use to analyze the system, represents a reasonable approximation of the real world traffic; and the use of a constant service time to represent the time spent by a vehicle driving across the intersection is also a fair approximation of the real world behaviour of drivers.

For $t \geq 0$, let $(L(t), \xi(t))$ denote the state of the system at instant t, with $L(t)$ representing the number of vehicles in the system (in brief, the queue length) at instant t and $\xi(t)$ the state of the signal (in brief, the phase) at the same instant, with the set of phases being $\{1, 2, \ldots, G + 1\}$, such that: phases $1, 2, \ldots, M$ correspond to the initial M time intervals of duration T of a green period, phases $M + 1, M + 2, \ldots, G$ correspond to the successive time intervals of duration T associated with extensions of a green period, and phase $G + 1$ corresponds to the red periods of duration RT. In addition, let τ_n denote the instant (of time) of occurrence of the n-th change of state in the phase process $(\xi(t))$, with $\tau_0 = 0$.

A careful analysis of the traffic process $\{(L(t), \xi(t))\}$ leads to the conclusion that it is a Markov regenerative process with state space $\mathbb{N} \times \{1, 2, \ldots, G + 1\}$; see, e.g., [9] for details on Markov regenerative processes. Moreover, by observing the process $\{(L(t), \xi(t))\}$ at times τ_n, we obtain the embedded Markov chain $\{X_n\}$, with $X_n = (L(\tau_n), \xi(\tau_n))$, $n \in \mathbb{N}$, denoting the state of the system immediately after the n-th phase change, being an $M/G/1$ type Markov chain, a type of chain that was investigated in detail in [14].

The Markov chain $\{X_n\}$ has state space $\mathbb{N} \times \{1, 2, \ldots, G + 1\}$ and transition probability matrix

$$Q = \begin{bmatrix} B_0' & A_1 & A_2 & \cdots \\ A_0' & A_1 & A_2 & \cdots \\ 0 & A_0 & A_1 & \cdots \\ \vdots & & \ddots & \ddots & \ddots \end{bmatrix}, \tag{1}$$

where the A_k, A_0', and B_0 are $(G+1) \times (G+1)$ nonnegative matrices. The entries of the matrices A_k are given by

$$(A_k)_{ij} = \begin{cases} e^{-\lambda T} \dfrac{(\lambda T)^k}{k!}, & i = 1, 2, \ldots, G, \, j = i + 1 \\ e^{-\lambda RT} \dfrac{(\lambda RT)^{k-1}}{(k-1)!}, & i = G + 1, j = 1, \, k \geq 1 \\ 0, & \text{otherwise,} \end{cases}$$

The two matrices A_0' and B_0' have similar forms but must be treated separately; in detail,

$$(B_0')_{ij} = \begin{cases} e^{-\lambda T}, & i = 1, 2, \ldots, M - 1, \, j = i + 1 \\ e^{-\lambda T}, & i = M, M + 1, \ldots, G, \, j = G + 1 \\ e^{-\lambda RT}, & i = G + 1, \, j = 1 \\ 0, & \text{otherwise,} \end{cases}$$

and $(A_0')_{i,i+1} = (B_0')_{i,i+1}$ for $i = 1, 2, \ldots, M - 1$, $(A_0')_{i,G+1} = (B_0')_{i,G+1}$ for $i = M, M + 1, \ldots, G + 1$, and all remaining entries of A_0' are 0.

Note that, for $k \geq 1$: $(A_k)_{i\,i+1}$, $1 \leq i \leq G$, denotes the probability that k vehicles arrive in a time interval, of duration T, elapsing from a transition to phase i to the next subsequent phase transition, to phase $i + 1$; conversely, $(A_k)_{G+1\,1}$ denotes the probability that $k - 1$ vehicles arrive in a time interval elapsing from a transition to phase $G + 1$, starting a red signal, to the subsequent phase transition, to phase 1 and starting a green signal. The particular shape of Q is intuitive; in particular, the need for the introduction of the blocks A_0' and B_0' in the first column of Q arises from the fact that the decision on whether an extension of the green signal will occur is exclusively determined by having vehicles waiting in line or not at the moment at which a decision on such extension needs to be made.

From the structure of the matrix Q in (1), it follows that the Markov chain $\{X_n\}$ is of $M/G/1$ type, and the invariant probability vector associated with the stochastic matrix Q can be computed using a procedure similar to the one described in [15] in case the stationarity condition $\lambda(G + R) < G$ is satisfied, as assumed in the rest of the paper.

To end the section, we let $\mathbf{u} = [u^{(0)} \, u^{(1)} \, u^{(2)} \ldots]$ denote the invariant probability vector associated with the stochastic matrix Q, an infinite row vector such that $u^{(k)} = [u_{k1} \, u_{k2} \ldots u_{k\,G+1}]$, $k \geq 0$, is an $(G + 1)$-dimension row vector and $\mathbf{u}Q = \mathbf{u}$, $\mathbf{u}\mathbf{1} = 1$, with $\mathbf{1}$ denoting a column vector of ones. Solving this equation for $\mathbf{u}$ involves using a recursive matrix formula that is nicely described in [17].

The element u_{ki} denotes the stationary probability that, at the beginning of a period in a phase, there are k vehicles in the system and the system is in phase i. As such, the stationary probability of the number of vehicles in the system at the beginning of a phase being equal to k is given by

$$u_{k\bullet} = \sum_{i=1}^{G+1} u_{ki}, \quad k \geq 0.$$

Then, if we let $\mathbf{r} = [r_1\, r_2 \ldots r_{G+1}]$ denote the stationary probability vector of the embedded phase process $\{\xi(\tau_n)\}$, we have $r_i = \sum_{j=0}^{\infty} u_{ji}$ since we may also view u_{ki} as the long-run fraction of phase transitions that lead to phase i with k vehicles staying in the system immediately after the phase transition, and r_i as the long-run fraction of phase transitions that lead to phase i.

3   Long-Run Properties of the Traffic Process

In this section we characterize the long-run properties of the Markov regenerative traffic process $\{(L(t), \xi(t))\}$. We first derive the long-run distribution of the number of vehicles in the system, in Theorem 1, and obtain an expression for the long-run mean number of vehicles in the system, in Theorem 2. After that, we give an expression for the long-run mean sojourn time of vehicles in the system. Finally, we present the long-run distribution of the number of extensions of the green period, along with its mean.

We first note that the long-run fraction of phase i intervals that are initiated with k vehicles in the system, denoted by π_{ki}, satisfies

$$\pi_{ki} = \frac{u_{ki}}{\sum_{j=0}^{\infty} u_{ji}} = \frac{u_{ki}}{r_i}. \tag{2}$$

Of particular relevance are the long-run (and stationary) distributions of the number of vehicles in the system at the beginning of green light periods, $\{\pi_{k1}\}_{k \geq 0}$, and at the beginning of red light periods $\{\pi_{k,G+1}\}_{k \geq 0}$. For later use, we let $\mathbb{E}[L_i]$ denote the long-run mean number of vehicles in the system immediately after a transition to phase i, i.e.,

$$\mathbb{E}[L_i] = \sum_{k=0}^{\infty} k\, \pi_{ki}. \tag{3}$$

We now address the long-run properties of the phase process $\{\xi(t)\}$. This is a semi-Markov process with embedded Markov chain at phase transition epochs $\{\xi_n\}$, such that the amount of time the process remains in phase i in each visit to the phase is the constant

$$T_i = \begin{cases} T, & i \neq G+1 \\ RT, & i = G+1. \end{cases}$$

Resorting to the theory of semi-Markov processes (see, e.g., [4], Theorem 4.6) we conclude that the long-run fraction of time the traffic process spends in phase i,

$$p_{\bullet i} = \lim_{t \to \infty} \left[\frac{1}{t} \int_0^t 1_{\{\xi(s)=i\}} \, ds \right],$$

can be written as $p_{\bullet i} = r_i T_i / \sum_{j=1}^{G+1} r_j T_j$, which reduces to

$$p_{\bullet i} = \begin{cases} \frac{r_i}{\sum_{j=1}^G r_j + R\,r_{G+1}}, & i \neq G+1 \\ \frac{R\,r_{G+1}}{\sum_{j=1}^G r_j + R\,r_{G+1}}, & i = G+1. \end{cases} \tag{4}$$

We next address the computation of the long-run distribution of the number of vehicles in the system, L. For that, we let p_{ki} denote the long-run fraction of time there are k vehicles in the system with the system being in phase i, i.e.,

$$p_{ki} = \lim_{t \to \infty} \left[\frac{1}{t} \int_0^t 1_{\{L(s)=k, \xi(s)=i\}} \, ds \right],$$

implying that $p_{\bullet i} = \sum_{k=0}^{\infty} p_{ki}$, for $i = 1, 2, \ldots, G+1$. The following theorem expresses how the $\{p_{ki}\}$ may be computed from the $\{u_{ki}\}$.

Theorem 1. *For $k \in \mathbb{N}$ and $i \in \{1, 2, \ldots, G+1\}$,*

$$p_{ki} = \frac{\sum_{j=0}^k u_{ji}\, \mu_{k-j}(i)}{\left(\sum_{j=1}^G r_j + R\,r_{G+1}\right) T}, \tag{5}$$

where $\mu_l(i)$, $l \in \mathbb{N}$, is given by

$$\mu_l(i) = \begin{cases} \frac{1}{\lambda}\left[1 - e^{-\lambda T} \sum_{m=0}^l \frac{(\lambda T)^m}{m!}\right], & i \neq G+1 \\ \frac{1}{\lambda}\left[1 - e^{-\lambda R T} \sum_{m=0}^l \frac{(\lambda R T)^m}{m!}\right], & i = G+1 \end{cases} \tag{6}$$

Proof. From the theory of Markov regenerative processes (see, e.g., [4], Theorem 4.7), the definition of p_{ki} and the structure of the traffic process $\{L(t), \xi(t)\}$, it follows that

$$p_{ki} = \frac{\sum_{j=0}^k u_{ji}\, \theta_{ji}(k)}{\sum_{l=1}^{G+1} r_l T_l},$$

with $\theta_{ji}(k)$ denoting the expected amount of time there are k vehicles in the system during an interval of time in phase i initiated with j vehicles in the system. From (2) and since $\sum_{l=1}^{G+1} r_l T_l = \left(\sum_{j=1}^G r_j + R\,r_{G+1}\right) T$, in order to prove the theorem it remains to show that the quantities $\theta_{ji}(k)$ are equal to the quantities $\mu_{k-j}(i)$ defined in (6). This follows, for $i \in \{1, 2, \ldots, G+1\}$ and $0 \leq j \leq k$, from the following set of equalities:

$$\theta_{ji}(k) = \mathbb{E}\left[\int_0^{T_j} \mathbf{1}_{\{L(t)=k|L(0)=j,\xi(0)=i\}}\,dt\right]$$

$$= \int_0^{T_j} P\left(L(t) = k|L(0) = j, \xi(0) = i\right)\,dt$$

$$= \begin{cases} \int_0^T e^{-\lambda t}\dfrac{(\lambda t)^{k-j}}{(k-j)!}dt\,, & i \neq G+1 \\ \int_0^{RT} e^{-\lambda t}\dfrac{(\lambda t)^{k-j}}{(k-j)!}dt\,, & i = G+1 \end{cases}$$

$$= \begin{cases} \frac{1}{\lambda}\left[1 - e^{-\lambda T}\sum_{m=0}^{k-j}\frac{(\lambda T)^m}{m!}\right], & i \neq G+1 \\ \frac{1}{\lambda}\left[1 - e^{-\lambda RT}\sum_{m=0}^{k-j}\frac{(\lambda RT)^m}{m!}\right], & i = G+1 \end{cases},$$

where the last equality may be obtained using induction on $k - j$ (see [10]). □

Let $p_{k\bullet}$ denote the long-run fraction of time there are k vehicles in the system,

$$p_{k\bullet} = \lim_{t\to\infty}\left[\frac{1}{t}\int_0^t \mathbf{1}_{\{L(s)=k\}}ds\right].$$

Then, as $p_{k\bullet} = \sum_{i=1}^{G+1} p_{ki}$, we conclude from Theorem 1 that for $k \in \mathbb{N}$,

$$p_{k\bullet} = \frac{\sum_{i=1}^{G+1}\sum_{j=0}^{k} u_{ji}\,\mu_{k-j}(i)}{\left(\sum_{j=1, j\neq G+1}^{G} r_j + Rr_{G+1}\right)T}, \tag{7}$$

with $\mu_{k-j}(i)$ given in (6).

The following theorem provides a formula for the long-run mean number of vehicles in the system.

Theorem 2. *The long-run mean number of vehicles in the system is given by*

$$\mathbb{E}[L] = \frac{\sum_{i=1}^{G+1} r_i\mathbb{E}[L_i]T_i}{\left(\sum_{j=1}^{G} r_j + Rr_{G+1}\right)T} + \frac{\lambda T}{2}\frac{\sum_{j=1}^{G} r_j + R^2 r_{G+1}}{\sum_{j=1}^{G} r_j + Rr_{G+1}}. \tag{8}$$

Proof. From the structure of the traffic process $\{L(t), \xi(t)\}$ and the fact that

$$\mathbb{E}[L] = \lim_{t\to\infty}\left[\frac{1}{t}\int_0^t \sum_{k=0}^{\infty} k\,\mathbf{1}_{\{L(s)=k\}}\,ds\right],$$

it follows from the theory of Markov regenerative processes (see, *e.g.*, [4], Theorem 4.7) that

$$\mathbb{E}[L] = \frac{\sum_{i=1}^{G+1}\sum_{k=0}^{\infty} u_{ki}\,\delta_{ki}}{\sum_{l=1}^{G+1} r_l T_l} \tag{9}$$

with

$$\delta_{ki} = \mathbb{E}\left[\int_0^{T_i} \sum_{l=0}^{\infty}(k+l)\mathbf{1}_{\{L(t)=(k+l)|L(0)=k,\xi(0)=i\}}\, dt\right].$$

This equality can also be written as

$$\delta_{ki} = \int_0^{T_i} \sum_{l=0}^{\infty}(k+l)P(L(t)=(k+l)|L(0)=k,\xi(0)=i)\, dt$$

$$= \int_0^{T_i} \sum_{l=0}^{\infty}(k+l)e^{-\lambda t}\frac{(\lambda t)^l}{l!}\, dt$$

$$= kT_i + \frac{\lambda T_i^2}{2}.$$

From (9), taking into account (2) and the fact that $\sum_{l=1}^{G+1} r_l T_l^2 = (\sum_{j=1}^{G} r_j + R^2 r_{G+1})T^2$, we have

$$\mathbb{E}[L] = \frac{\sum_{i=1}^{G+1} r_i \sum_{k=0}^{\infty} k\pi_{ki}T_i}{(\sum_{j=1}^{G} r_j + Rr_{G+1})T} + \frac{\lambda}{2}\frac{\left(\sum_{j=1}^{G} r_j + R^2 r_{G+1}\right)T^2}{(\sum_{j=1}^{G} r_j + Rr_{G+1})T}.$$

The expression (8) for $\mathbb{E}[L]$ now follows since $\sum_{k=0}^{\infty} k\pi_{ki} = \mathbb{E}[L_i]$. $\qquad\square$

When assessing traffic systems, delay of vehicles is a major concern. The long-run distribution of the sojourn time of a vehicle in the system is complex, but can be derived following a procedure similar to the one used in Sect. 4 of [15], with the necessary adaptations. One immediate contribution can be put in terms of the computation of the long-run mean sojourn time of a vehicle in the system, $\mathbb{E}[W]$. According to our model, it can be derived from Little's formula (cf. for instance [9]) applied to expression (8), giving:

$$\mathbb{E}[W] = \frac{\mathbb{E}[L]}{\lambda} = \frac{\sum_{i=1}^{G+1} r_i \mathbb{E}[L_i]T_i}{\lambda T\left(\sum_{j=1}^{G} r_j + Rr_{G+1}\right)} + \frac{T}{2}\frac{\sum_{j=1}^{G} r_j + R^2 r_{G+1}}{\sum_{j=1}^{G} r_j + Rr_{G+1}}. \tag{10}$$

The setting considered in this paper allows extensions of the green signal, which occur when there are cars waiting to be served at the end of the minimum duration of a green period. An important measure is the long-run mean number of extensions (or equivalently the time of extension) of a green period, which is clearly not constant as it is the case in a non-actuated signalized intersection.

Let us consider a random variable N_G whose distribution is the long-run distribution of the number of extensions of the green period. By establishing that $P(N_G > k-1) = \frac{r_{M+k}}{r_M}$, for $k = 1, 2, \ldots, G-M$, one can conclude that the long run fraction of green periods with k extensions is

$$P(N_G = k) = \begin{cases} \dfrac{r_{M+k} - r_{M+k+1}}{r_M}, & k = 0, 1, \ldots, G - M - 1 \\[2ex] \dfrac{r_G}{r_M}, & k = G - M \end{cases} \tag{11}$$

and the long-run mean number of extensions of the green period is

$$\mathbb{E}[N_G] = \sum_{k=1}^{G-M} \frac{r_{M+k}}{r_M}. \tag{12}$$

4   Case Study

In order to illustrate the applicability of the formulation that we propose, we consider an intersection with 3 traffic streams having a primary phase and a secondary phase as illustrated in Fig. 1. The primary phase, associated to the two main traffic streams, is not actuated. A sensor is placed two meters before the stop line on the secondary street and the control of the secondary phase, associated to this street, is actuated by means of the information provided by the sensor (inter-arrival times). The time plan is the following: $M = 4$, $G = 15$, $R = 15$. We consider $T = 2\,\mathrm{s}$. With this time plan, the maximum duration of extended green is $T(G - M) = 22\,\mathrm{s}$. The vehicle arrival rate on the main street is assumed to be 800 veh/hour for each stream. The performance measures that we present correspond only to the actuated stream. Note that, in this situation, the vehicle arrival rates on the main street do not influence the measures on the secondary street.

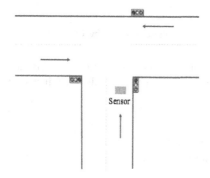

Fig. 1. Scheme of the intersection, indicating the two phases. (Color figure online)

Regarding the microsimulator, the following set up was used (see [19] for details):

- vehicle's characteristics: desired speed - Gaussian $(13.9\,\mathrm{m/s}, 0.2\,\mathrm{m/s})$; maximum acceleration - Gaussian $(1.7\,\mathrm{m/s^2}, 0.3\,\mathrm{m/s^2})$; length of a vehicle - Gaussian $(4.0\,\mathrm{m}, 0.3\,\mathrm{m})$;
- number of replications: between 100 and 1000, depending on arrival rate, controlling for the standard deviation of the Monte-Carlo error to be smaller than 1;

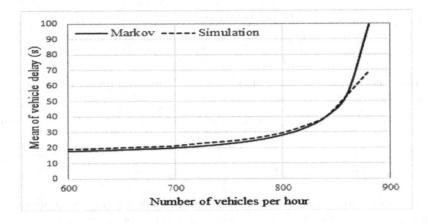

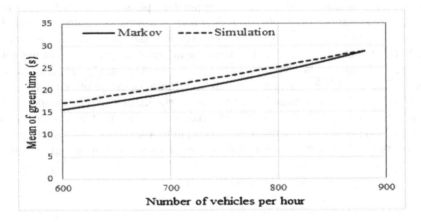

Fig. 2. Comparison between estimates provided by the Markov chain based model and by the microsimulation model: long-run mean delay of drivers (top); and long-run mean queue length (bottom).

– warm up time: 600 s;
– run time: 2 h/replica.

Vehicle's characteristics have been set on the basis of information collected concerning the real operations of traffic in urban areas (see [19]). In the simulator, vehicles move according to a car-following model, that is, essentially drivers adapt the speed of their vehicles to that of the vehicle in front of them, so that their heading is kept above a minimum value which corresponds to the drivers perception of safety (see, e.g. [16] for a review of car-following models). This level of detail in the description of the behaviour of vehicles, which is typical of micro-simulation models, is not possible in the Markov model that we propose.

Figure 2 shows estimates of the long-run mean waiting time of drivers and the long-run mean duration of the green signal obtained by the model presented in

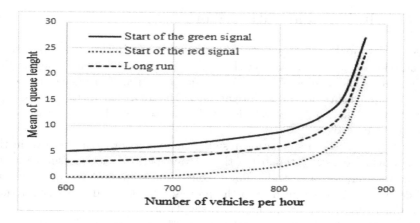

Fig. 3. Estimates of the long-run mean queue length provided by the Markov chain based model: long-run mean queue length; long-run mean queue length at the start of the green signal; long-run mean queue length at the start of the red signal (overflow queue).

the previous sections together with the results obtained by using the simulation model described in [19], considering different vehicle arrival rates on the secondary street. We use the word "Markov" in the figures to refer to the proposed model.

The long-run mean queue length in depicted in Fig. 3, along with the long-run mean queue length at two different time points that are of interest in the signal cycle, namely at the start of the green signal and at the start of the red signal.

We can see the exponential increase of the mean waiting time when the vehicle arrival rate increases, as expected. The results suggest that, from moderate values of the vehicle arrival rate to considerable higher values (but away from the saturation level) the estimates of the mean delay of drivers given by the Markov based model through expressions (8)–(10) are quite close to the simulation results. Unfortunately the approximation is not so good when we consider very large vehicle arrival rates (i.e. close to the saturation level). This fact may be explained by the diversity of reactions that are typical of drivers' behaviour and of interactions between vehicles which is mimicked in the simulation model quite closely (cf. [19]) but is hardly taken into account in a Markov or renewal type process modelling. For instance, drivers may decelerate promptly when approaching a slowing vehicle or queue. Interactions between vehicles have a major impact when system parameters are close to the boundary of the stationarity region of the traffic system. We can also observe the exponential increase of the queues when the vehicle arrival rate increases, as expected, and an increasing mean duration of the green period due to the occurrence of several extensions of the green period becoming common.

5 Conclusions and Future Work

A detailed probabilistic description of the delay of vehicles in semi-actuated signalized traffic intersections, as well as of the length of queues and the duration of the green signal can be obtained by considering an $M/D/1$ queue with server vacations and using, for its investigation, a Markov-regenerative process that keeps track of the number of vehicles at the intersection along the phase of the signal cycle over time.

When compared to simulation results, the expressions that we give in the paper provide realistic estimates of the relevant performance measures investigated. However, for large traffic flows (congestion scenarios) the queue length and delay measures obtained from the proposed model tend to be larger than the estimates returned by the numerical simulator.

Future work will address the extension of the analysis for the case of semi-actuated control in which extensions are also allowed for the red signal.

Acknowledgments. The first author was partially supported by CMUP under a grant of the project UID/MAT/00144/2013, financed by FCT/MEC (PIDDAC). This research was partially supported by CMUP (UID/MAT/00144/2013) and CEMAT (UID/Multi/04621/2013), funded by FCT (Portugal) with National (MEC) and European structural funds through the programs FEDER, under partnership agreement PT2020.

References

1. Al Hanbali, A., de Haan, R., Boucherie, R.J., van Ommeren, J.: Time-limited polling systems with batch arrivals and phase-type service times. Ann. Oper. Res. **198**(1), 57–82 (2012)
2. Brockfeld, E., Wagner, P.: Validating microscopic traffic flow models. In: Intelligent Transportation Systems Conference, ITSC 2006, pp. 1604–1608. IEEE (2006)
3. Doshi, B.T.: Queueing systems with vacations - a survey. Queueing Syst. **1**(1), 29–66 (1986)
4. El-Taha, M., Stidham Jr., S.: Sample-path analysis of queueing systems, vol. 11. Springer Science & Business Media, Berlin (2012)
5. Frigui, I., Alfa, A.: Analysis of a time-limited polling system. Comput. Commun. **21**(6), 558–571 (1998)
6. de Haan, R., Boucherie, R.J., van Ommeren, J.: A polling model with an autonomous server. Queueing Syst. **62**(3), 279–308 (2009)
7. Heidemann, D.: Queue length and delay distributions at traffic signals. Transp. Res. Part B: Methodol. **28**(5), 377–389 (1994)
8. Hu, X., Tang, L., Ong, H.: A $M/D^X/1$ vacation queue model for a signalized intersection. Comput. Ind. Eng. **33**(3), 801–804 (1997)
9. Kulkarni, V.: Modeling and Analysis of Stochastic Systems. Chapman & Hall/CRC Texts in Statistical Science. Taylor & Francis, Abingdon (1996). http://books.google.ch/books?id=HOPxhUonodgC
10. Kwiatkowska, M., Norman, G., Pacheco, A.: Model checking expected time and expected reward formulae with random time bounds. Comput. Math. Appl. **51**(2), 305–316 (2006)

11. Leung, K.K.: Cyclic-service systems with nonpreemptive, time-limited service. IEEE Trans. Commun. **42**(8), 2521–2524 (1994)
12. Leung, K.K., Eisenberg, M.: A single-server queue with vacations and non-gated time-limited service. Perform. Eval. **12**(2), 115–125 (1991)
13. Lin, D., Wu, N., Zong, T., Mao, D.: Modeling the impact of side-street traffic volume on major-street green time at isolated semi-actuated intersections for signal coordination decisions. In: Transportation Research Board 95th Annual Meeting, pp. 16–29 (2016)
14. Neuts, M.F.: Structured Stochastic Matrices of M/G/1 Type and Their Applications. Marcel Dekker Inc., New York (1989)
15. Pacheco, A., Simões, M.L., Milheiro-Oliveira, P.: Queues with server vacations as a model for pretimed signalized urban traffic. Transp. Sci. (2017, in press)
16. Panwai, S., Dia, H.: Comparative evaluation of microscopic car-following behavior. IEEE Trans. Intell. Transp. Syst. **6**(3), 314–325 (2005)
17. Ramaswami, V.: A stable recursion for the steady state vector in markov chains of $M/G/1$ type. Stoch. Models **4**(1), 183–188 (1988)
18. Ryus, P., Vandehey, M., Elefteriadou, L., Dowling, R.G., Ostrom, B.K.: Highway capacity manual 2010. Tr News **273**, 45–48 (2011)
19. Simões, M.L., Milheiro-Oliveira, P., Pires da Costa, A.: Modeling and simulation of traffic movements at semiactuated signalized intersections. J. Transp. Eng. **136**(6), 554–564 (2009)
20. Sun, B., Wu, N., Ge, Y.E., Kim, T., Zhang, H.M.: A new car-following model considering acceleration of lead vehicle. Transport **31**(1), 1–10 (2016)
21. Takagi, H.: Analysis and application of polling models. In: Haring, G., Lindemann, C., Reiser, M. (eds.) Performance Evaluation: Origins and Directions. LNCS, vol. 1769, pp. 423–442. Springer, Heidelberg (2000). doi:10.1007/3-540-46506-5_18
22. Vishnevskii, V., Semenova, O.: Mathematical methods to study the polling systems. Autom. Remote Control **67**(2), 3–56 (2006)
23. Viti, F., Van Zuylen, H.J.: The dynamics and the uncertainty of queues at fixed and actuated controls: a probabilistic approach. J. Intell. Transp. Syst. **13**(1), 39–51 (2009)
24. Viti, F., Van Zuylen, H.J.: Probabilistic models for queues at fixed control signals. Transp. Res. Part B: Methodol. **44**(1), 120–135 (2010)
25. Webster, F.V.: Traffic signal settings. Technical report no. 39. Road Research Laboratory, HMSO, London (1958)

Aggregation and Truncation of Reversible Markov Chains Modulo State Renaming

Andrea Marin$^{(\boxtimes)}$ and Sabina Rossi

DAIS - Università Ca' Foscari, Venezia, Italy
{marin,sabina.rossi}@unive.it

Abstract. The theory of time-reversibility has been widely used to derive the expressions of the invariant measures and, consequently, of the equilibrium distributions for a large class of Markov chains which found applications in optimisation problems, computer science, physics, and bioinformatics. One of the key-properties of reversible models is that the truncation of a reversible Markov chain is still reversible. In this work we consider a more general notion of reversibility, i.e., the reversibility modulo state renaming, called ρ-reversibility, and show that some of the properties of reversible chains cannot be straightforwardly extended to ρ-reversible ones. Among these properties, we show that in general the truncation of the state space of a ρ-reversible chain is not ρ-reversible. Hence, we derive further conditions that allow the formulation of the well-known properties of reversible chains for ρ-reversible Markov chains. Finally, we study the properties of the state aggregation in ρ-reversible chains and prove that there always exists a state aggregation that associates a ρ-reversible process with a reversible one.

1 Introduction

Reversibility of Markov chains at discrete or continuous time has been extensively studied in [13,25]. Given a stationary Markov chain $X(t)$ we say that it is reversible if for all $t_1, t_2, \ldots, t_n, \tau$, $(X(t_1), \ldots, X(t_n))$ has the same equilibrium distribution as $(\rho(X)(\tau - t_1), \ldots, \rho(X)(\tau - t_n))$ where $t_1, \ldots, t_n, \tau$ belongs to the time domain, i.e., $\mathbb{Z}$ for discrete time Markov chains (DTMCs) and $\mathbb{R}$ for continuous time Markov chains (CTMCs). Reversibility is a key-property for studying the stationary behaviour of Markov chains and there are several examples of models with underlying reversible processes such as the *loss networks* [14] which found applications for studying telecommunication systems, models of wireless networks [5] just to mention a non exhaustive list of applications. In many practical cases, reversibility allows for the derivation of an exact analysis of the stationary behaviour of the model without resorting to simulation, approximate decompositions (see e.g., [3,6]) or limit-based analysis (see e.g., [4,7]).

However, the largest application field of reversible Markov chains is in queueing theory. Queueing theory is the foundation of many works in operation research (see, e.g., [8,16,23] just to mention some recent works) and some of them are based on reversible models or their variation [1,2,12,13,24].

© Springer International Publishing AG 2017
N. Thomas and M. Forshaw (Eds.): ASMTA 2017, LNCS 10378, pp. 152–165, 2017.
DOI: 10.1007/978-3-319-61428-1_11

Markov chain reversibility is a special case of a more general notion of reversibility that we call ρ-reversibility. A ρ-reversible chain $X(t)$ is stochastically indistinguishable from $X(\tau - t)$ modulo a state renaming which is a bijective function ρ from the chain's state space S to itself. An example of such a chain is shown in Fig. 1 where we can easily see that the forward CTMC (Fig. 1-(a)) is not reversible since a simple necessary structural condition for reversibility is that whenever there is a transition from state s to state s' there is also its inverse from s' to s. Figure 1-(b) shows the transition diagrams of $X(\tau - t)$ and we can observe that it is stochastically indistinguishable from $X(t)$ modulo the renaming of states $\rho(1) = 2, \rho(2) = 1, \rho(3) = 4$ and $\rho(4) = 3$.

(a)   (b)

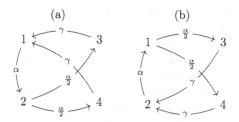

Fig. 1. A simple ρ-reversible CTMC: (a) Forward process, (b) Reversed process.

In this case function ρ is an involution. In the literature of stochastic processes when ρ is an involution the notion of ρ-reversibility is known as *dynamic reversibility* and has been studied in [13,25]. It is worth of notice that the concept of ρ-reversibility is more general than that of dynamic reversibility, i.e., there exist Markov processes which are ρ-reversible but there does not exist any involution for which they are also dynamically reversible [18,21].

Reversible Markov chains enjoy some important properties that can be readily formulated also for ρ-reversible chains. Specifically, in both cases one may decide if a chain is reversible/ρ-reversible by inspection of a base of minimal cycles of the chain and the computation of a non-trivial invariant measure can be done by performing only multiplications and using the detailed balance equations [13,19,21,25]. However, other important properties that hold for reversible Markov chains do not straightforwardly hold for the ρ-reversible ones. Specifically, if S is the state space of a reversible CTMC, $A \subset S$ and if the graph of A is irreducible, then also the chain whose state space is A and the transitions are only those of the original one for the states in A is reversible. We say in this case that the resulting process is *truncated* to the set A. A similar result holds if the transition rates from set $S \setminus A$ to A are changed by the same multiplicative factor. In this paper we prove that, in general, these results do not hold for ρ-reversible and dynamically reversible chains, but they require some further conditions that are trivially satisfied in the case of reversible chains. It is worth to observe that, to the best of our knowledge, this is the first work that studies the truncation properties for Markov chains that are reversible modulo state renaming, including those that are know to be dynamically reversible. In fact, in [13,25] the authors consider only the truncation of reversible processes. We also investigate the definition of the aggregated process for reversible and ρ-reversible chains and prove that they also are reversible or ρ-reversible.

The paper is structured as follows. Section 2 illustrates the preliminary notions and the notation which are necessary to keep the paper self-contained. In Sects. 3 and 4 we prove the new results about process aggregation and truncation, respectively. Finally, Sect. 5 concludes the paper.

2   Preliminaries

Let us consider a Markov chain $X(t)$ defined on the state space $\mathcal{S}$. For the sake of brevity we study the continuous time case, i.e., $t \in \mathbb{R}$. Given a stationary CTMC $X(t)$ the process $X(\tau - t)$, denoted by $X^R(t)$, is still a stationary CTMC [13] and the equilibrium state probability π for $X(t)$ is the same of that of $X^R(t)$.

In [19] the notion of reversibility for CTMC has been generalized to a notion of reversibility under state renaming named ρ-reversibility.

Formally, a *renaming* ρ over the state space of a Markov chain is a bijection from $\mathcal{S}$ to itself. For a Markov chain $X(t)$ with state space $\mathcal{S}$ we denote by $\rho(X)(t)$ the same process where the state names are changed according to ρ.

The notion of ρ-reversibility is defined as follows.

Definition 1 *(ρ-reversibility)* [19,21]. *Let $X(t)$ be a stationary CTMC with state space $\mathcal{S}$ and ρ be a renaming on $\mathcal{S}$. $X(t)$ is said to be ρ-reversible if for all $t_1, t_2, \ldots, t_n, \tau \in \mathbb{R}$, $(X(t_1), \ldots, X(t_n))$ has the same equilibrium distribution as $(\rho(X)(\tau - t_1), \ldots, \rho(X)(\tau - t_n))$. Moreover, if ρ is the identity we say that $X(t)$ is reversible whereas if ρ is a non-trivial involution, i.e., $\forall s \in \mathcal{S} \, \rho(\rho(s)) = s$ but ρ is not the identity, then $X(t)$ is said to be dynamically reversible [13].*

Notice that from Definition 1 and the fact that $X(t)$ and $X^R(t)$ have the same equilibrium state distribution, it follows that:

$$\pi(s) = \pi(\rho(s)) \quad \text{for all } s \in \mathcal{S}.$$

It is important to observe that a CTMC may be ρ_1-reversible and ρ_2-reversible for some $\rho_1 \neq \rho_2$. In [18] we prove that the extension of dynamic reversibility to ρ-reversibility is non-trivial since there exist CTMCs such that they have a function ρ for which they are ρ-reversible but there does not exist any involution that makes them dynamically reversible.

The following proposition, proved in [19], gives necessary and sufficient conditions for a CTMC to be ρ-reversible given a certain ρ.

Proposition 1 *(ρ-detailed balance equations). Let $X(t)$ be an ergodic CTMC with state space $\mathcal{S}$ and infinitesimal generator matrix $\mathbf{Q}$. Let ρ be a renaming on $\mathcal{S}$. $X(t)$ is ρ-reversible if and only if there exists a collection of positive numbers*

$\pi(s)$, $s \in \mathcal{S}$, summing to unity that satisfy the following system of ρ-detailed balance equations:

$$\pi(s)q(s, s') = \pi(\rho(s'))q(\rho(s'), \rho(s)) \quad \text{for all } s, s' \in \mathcal{S}, \tag{1}$$

where $q(s, s')$ denotes the transition rate from state s to s', with $s \neq s'$. If such a solution $\boldsymbol{\pi}$ exists then it is the equilibrium distribution of both $X(t)$ and $\rho(X^R)(t)$ and $\pi(s) = \pi(\rho(s))$ for all $s \in \mathcal{S}$.

If the equilibrium distribution of $X(t)$ is known, the following corollary gives a straightforward way to decide if $X(t)$ is ρ-reversible given a certain ρ.

Corollary 1. *Let $X(t)$ be an ergodic CTMC with state space $\mathcal{S}$, infinitesimal generator matrix $\mathbf{Q}$ and equilibrium distribution $\boldsymbol{\pi}$. Let ρ be a renaming on the state space $\mathcal{S}$. If the transition rates of $X(t)$ satisfy the following system of equations:*

$$\pi(s)q(s, s') = \pi(s')q(\rho(s'), \rho(s)) \quad for \ all \ s, s' \in \mathcal{S}$$

then $X(t)$ is ρ-reversible.

The previous methods to decide the property of ρ reversibility are based on the computation or the knowledge of the equilibrium distribution. In contrast, Kolmogorov's critera are purely structural, i.e., they depend only on the structure of the underlying transition graph and on the transition rates and do not require the solution of a linear system of equations.

Proposition 2. *Let $X(t)$ be an ergodic CTMC with state space $\mathcal{S}$ and infinitesimal generator matrix $\mathbf{Q}$, and ρ be a renaming on $\mathcal{S}$. $X(t)$ is ρ-reversible if and only if for every finite sequence $s_1, s_2, \ldots s_n \in \mathcal{S}$,*

$$q(s_1, s_2) \ \cdots \ q(s_{n-1}, s_n) \, q(s_n, s_1) =$$
$$q(\rho(s_1), \rho(s_n)) \, q(\rho(s_n), \rho(s_{n-1})) \ \cdots \ q(\rho(s_2), \rho(s_1))$$

and $q(s) = q(\rho(s))$ for every state $s \in \mathcal{S}$.

The equilibrium distribution of a ρ-reversible CTMC can be computed as stated in Proposition 3. Notice that Proposition 3 gives a numerically stable method to compute a non-trivial invariant measure of the process since for each state it requires the computation only of products.

Proposition 3. *Let $X(t)$ be an ergodic CTMC with state space $\mathcal{S}$ and infinitesimal generator matrix $\mathbf{Q}$, ρ be a renaming on $\mathcal{S}$, and $s_0, s_1, s_2, \ldots s_n = s \in \mathcal{S}$ be a finite sequence of states. If $X(t)$ is ρ-reversible then for all $s \in \mathcal{S}$,*

$$\pi(s) = C_{s_0} \prod_{k=1}^{n} \frac{q(\rho(s_{k-1}), \rho(s_k))}{q(s_k, s_{k-1})} \tag{2}$$

where $s_0 \in \mathcal{S}$ is an arbitrary reference state and $C_{s_0} \in \mathbb{R}^+$.

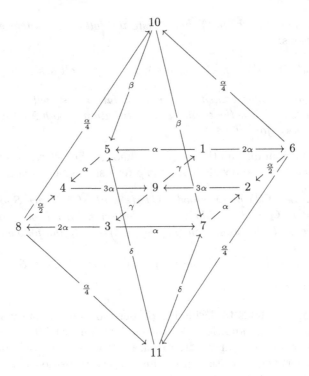

Fig. 2. A ρ-reversible CTMC.

Recall that a permutation ρ on a set S admits a unique decomposition into *cycles* of different states:

$$(s, \rho(s), \rho(\rho(s)), \ldots, \rho^n(s) \equiv s).$$

The set of states in a cycle form an orbit. Then, every permutation can be decomposed into a collection of cycles on disjoint orbits.

Example 1. If we consider the CTMC depicted in Fig. 2 we can prove that it is ρ-reversible where ρ is described by the following orbits: (1, 2, 3, 4); (5, 6, 7, 8); (9); (10); (11).

Now, we review an aggregation technique for CTMCs that preserve the equilibrium distribution, i.e., the equilibrium probability of the macro state is given by the sum of the equilibrium probability of its elements in the original, non aggregated, process. More formally, let $\sim$ be an equivalence relation over the state space S of a CTMC $X(t)$. In general, the process obtained by the observation of the macro state jump process is *not* a Markov process (for instance the residence time in an aggregated state is not exponentially distributed) unless we have a lumping [15]. However, we can still define CTMC $\widetilde{X}(t)$ corresponding to a certain aggregation $\sim$ as follows: the state space is the set of the equivalence classes $\mathcal{S}/\sim$ and its infinitesimal generator matrix $\widetilde{\mathbf{Q}}$ can be derived from the following general aggregation equation for any $S_i, S_j \in \mathcal{S}/\sim$,

$$\widetilde{q}(S_i, S_j) = \frac{\sum_{s' \in S_i} \pi(s') \sum_{s \in S_j} q(s', s)}{\sum_{s' \in S_i} \pi(s')} \tag{3}$$

The following proposition shows that the equilibrium distribution of the aggregated process is such that the equilibrium probability of each macro-state is the sum of the equilibrium probabilities of the states in the original process forming it.

Proposition 4. *Let $X(t)$ be an ergodic CTMC with state space $\mathcal{S}$ and $\sim$ be an equivalence relation over $\mathcal{S}$. Let $\widetilde{X}(t)$ be the aggregated process with respect to $\sim$. Let $\boldsymbol{\pi}$ and $\widetilde{\boldsymbol{\pi}}$ be the equilibrium distributions of $X(t)$ and $\widetilde{X}(t)$, respectively. Then for all $S \in \mathcal{S}/\sim$,*

$$\widetilde{\pi}(S) = \sum_{s \in S} \pi(s).$$

3   Aggregation of ρ-reversible Processes

Aggregation is a technique for reducing the state space of a model and hence for deriving some quantitative measures more efficiently. Unfortunately, for general processes, an aggregation of states that respects the equilibrium distributions (i.e., the equilibrium probability of a macro state is given by the sum of the equilibrium probabilities of the states that it aggregates) is as hard to compute as the computation of the model equilibrium distribution as shown by Eq. (3). Strong lumping [15] is a structural approach to state aggregation, i.e., the definition of the aggregated chain does not require the knowledge of its equilibrium distribution. In this section we will show that also the class of ρ-reversible CTMCs can be aggregated in a process whose transition rates can be obtained without the knowledge of the equilibrium distribution and hence can be performed efficiently. Before stating the results on the aggregation in ρ-reversible (and hence also reversible) CTMCs, we need to introduce a definition of compatibility of an aggregation with a renaming ρ. Intuitively, we say that an aggregation $\sim$ respects renaming ρ if its equivalence classes are either singletons or if they contain more states then they must cluster together all the states of the corresponding orbits.

Definition 2. *An aggregation $\sim$ respects a renaming ρ on $\mathcal{S}$ if for each $S \in \mathcal{S}/\sim$ at least one of the following conditions is satisfied:*

- *$|S| = 1$, or*
- *$s \in S$ implies $\rho(s) \in S$.*

We stress on the fact that Definition 2 does not require that the state partitions correspond to the orbits of ρ, but it states that if we aggregate two states, then all the states in their orbits must belong to the same partition. However, states that are not aggregated do not need to satisfy this conditions.

Example 2. Consider the CTMC with states space $S = \{s_1, \ldots s_8\}$ and let the orbit of ρ be $(s_1, s_2), (s_3, s_4), (s_5, \ldots, s_8)$, then the following partitions of states respects ρ:

- $S_1 = \{s_1, s_2\}$, $S_2 = \{s_3, \ldots, s_8\}$
- $S_1 = \{s_1\}$, $S_2 = \{s_2\}$, $S_3 = \{s_3, s_4\}$, $S_4 = \{s_5, \ldots, s_8\}$
- $S_i = \{s_i\}$ (the trivial partition)

Theorem 1 states that an aggregation $\sim$ of a ρ-reversible chain $X(t)$ is $\tilde{\rho}$-reversible for a certain renaming $\tilde{\rho}$ if $\sim$ respects ρ.

Theorem 1. *Let $X(t)$ be a ρ-reversible CTMC and let $\sim$ be an aggregation that respects ρ according to Definition 2. Then, Markov chain $\tilde{X}(t)$ is $\tilde{\rho}$-reversible where $\tilde{\rho}$ is defined as follows:*

$$\tilde{\rho}(S_i) = \begin{cases} S_j & \text{if } S_i = \{s\} \wedge S_j = \{\rho(s)\}, \\ S_i & \text{if } |S_i| > 1. \end{cases} \tag{4}$$

Let us analyse some consequences of Theorem 1. Let $X(t)$ be a ρ-reversible CTMC with state space S and $\sim$ be the equivalence relation over S such that $s_1 \sim s_2$ if and only if s_1 and s_2 belongs to the same orbit with respect to the permutation ρ. Then clearly $\sim$ respects ρ according to Definition 2, $\tilde{\rho}$ is the identity on $S/\sim$ and $S/\sim$ denotes the set of all orbits induced by ρ in S. In this case we say that $\sim$ is the equivalence relation induced by ρ in S.

Corollary 2. *Let $X(t)$ be a ρ-reversible CTMC with state space S and infinitesimal generator matrix $\mathbf{Q}$. Let $\sim$ be the equivalence relation over S induced by ρ. Then $\tilde{X}(t)$ is reversible.*

Proof. The proof follows from Theorem 1 and the observation that $\tilde{\rho}$ is the identity (see Definition 1). $\square$

For this type of aggregation the transition rates of the aggregated process can be calculated without the computation of the equilibrium state distribution π.

Proposition 5. *Let $X(t)$ be a ρ-reversible CTMC with state space S and infinitesimal generator matrix $\mathbf{Q}$. Let $\sim$ be the equivalence relation over S induced by ρ. Then, the infinitesimal generator matrix $\tilde{\mathbf{Q}}$ of $\tilde{X}(t)$ is defined as:*

$$\tilde{q}(S_i, S_j) = \frac{\sum_{s' \in S_i} \sum_{s \in S_j} q(s', s)}{|S_i|} \tag{5}$$

where $|S_i|$ denotes the cardinality of the orbit S_i.

The next corollary follows immediately from Theorem 1 and states that any aggregation of a reversible chain is still reversible.

Corollary 3. *Let $X(t)$ be a reversible CTMC, then for any aggregation $\sim$ on its state space S we have that $\tilde{X}(t)$ is still reversible.*

Proof. Observe that if $X(t)$ is reversible then ρ is the identity and hence any aggregation $\sim$ respects ρ. The proof follows by observing that by definition also $\widetilde{\rho}$ is the identity and hence $\widetilde{X}(t)$ is reversible.   □

Example 3. Let us aggregate the ρ-reversible process of Fig. 2 with respect to relation $\sim$ induced by the orbits of the CTMC. Then, by Proposition 5 we can straightforwardly derive the aggregated process of Fig. 3. It is easy to observe that the resulting CTMC is reversible.

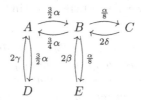

Fig. 3. Aggregation according to the orbits of the CTMC shown in Fig. 2.

4   Truncation of ρ-reversible Processes

The truncation of a reversible CTMC is a very useful technique to study models in which some agents compete for a set of resources. For instance, consider a reversible chain that models N agents performing a set of operations some of which require a certain resource whose availability is $M < N$. In order to study the equilibrium properties of the model, we may assume that the resource is always available for all the agents and prove the reversibility of the underlying process, then we have to exclude the transitions that would take the model to states in which more than M resources are used simultaneously. In [13, Lemma 1.9, Corollary 1.10] the author proves that if the original process is reversible then also the truncated one is reversible.

In this section we study the same problem with ρ-reversible processes. The main result that we derive is that, in general, the truncation of a ρ-reversible process is not ρ-reversible. In fact, in order to prove the analogue result of Lemma 1.9 and Corollary 1.10 of [13] we require that the truncation respects the orbits of ρ, i.e., each orbit is either entirely truncated or kept.

A reversible CTMC may be altered by changing the transition rates in such a way that the equilibrium distribution is not changed. As observed in [13] if a reversible CTMC $X(t)$ has transition rate $q(s,u) > 0$ and $q(u,s) > 0$, then also the CTMC $X'(t)$ whose transition rates are the same of $X(t)$ with the exception of $q'(s,u) = cq(s,u)$ and $q'(u,s) = cq(u,s)$ for $c > 0$ is still reversible. The result follows immediately from Proposition 1 assuming ρ to be the identity. We notice that this result is in general not applicable to ρ-reversible CTMCs since the modification of $q(s,u)$ to $cq(s,u)$ changes the residence time of state i and the definition of ρ-reversibility requires that all the states in the orbit of s must have the same residence time.

Example 4. Let us consider the model of Fig. 1-(a) and let us write the ρ-detailed balance equation associated with the transition from state 1 to state 2:

$$\pi(1)q(1,2) = \pi(\rho(2))q(\rho(2), \rho(1)).$$

Notice that, since $\rho(1) = 2$ and $\rho(2) = 1$ we have that $q(1,2) = q(\rho(2), \rho(1))$, hence the detailed balance equation is satisfied even if we set $q'(1,2) = c\alpha$, for $c > 0$ and $c \neq 1$. Nevertheless, the CTMC $X'(t)$ is *not* ρ-reversible since the residence time in state 1 has mean $(c\alpha)^{-1}$ while in state 2 has mean α^{-1}.

The following Lemma is the version of Lemma 1.9 in [13] for ρ-reversible CTMCs.

Lemma 1. *Let $X(t)$ be a ρ-reversible CTMC with state space $\mathcal{S}$ and let $\sim$ be an equivalence relation that induces only two non-empty equivalence classes $\mathcal{A} \subset \mathcal{S}$ and $\mathcal{S} \setminus \mathcal{A}$. Then, if $\sim$ respects ρ we have that for any positive constant $c \in \mathbb{R}$ the chain $X'(t)$ whose transition rates $q'(s, u)$ are defined as follows:*

$$q'(s, u) = \begin{cases} cq(s, u) & \text{if } s \in \mathcal{A} \wedge u \in \mathcal{S} \setminus \mathcal{A} \\ q(s, u) & \text{otherwise} \end{cases}$$

is still ρ-reversible if the residence time of the states in $X'(t)$ are identically distributed for all the states belonging to the same orbit. Moreover, if $X'(t)$ is ρ-reversible, then the equilibrium distribution $\pi'(s)$ for $X'(t)$ is:

$$\pi'(s) = \begin{cases} B\pi(s) & \text{if } s \in \mathcal{A} \\ Bc\pi(s) & \text{if } s \in \mathcal{S} \setminus \mathcal{A} \end{cases},$$

where B is a normalising constant.

The following corollary follows from Lemma 1 where $c = 0$ and is the analogue of Corollary 1.10 in [13].

Corollary 4. *Let $X(t)$ be a ρ-reversible CTMC with state space $\mathcal{S}$ and let $\sim$ be an equivalence relation that induces only two non-empty equivalence classes $\mathcal{A} \subset \mathcal{S}$ and $\mathcal{S} \setminus \mathcal{A}$. Let $\sim$ respect ρ, and define the chain $X'(t)$ on the state space $\mathcal{A}$ with transition rates:*

$$q'(s, u) = \begin{cases} q(s, u) & \text{if } s, u \in \mathcal{A} \\ 0 & \text{otherwise} \end{cases}.$$

Then if $X'(t)$ is irreducible and the residence time of every state $s \in \mathcal{A}$ is the same of every other state u in the same orbit of s, we have that $X'(t)$ is ρ-reversible. In this case the equilibrium probabilities of $s \in \mathcal{A}$ are:

$$\pi(s) = \frac{\pi(s)}{\sum_{u \in \mathcal{S}} \pi(u)}.$$

Proof. The proof follows straightforwardly from Lemma 1.   □

Example 5. Let us consider a manufacturing system where K independent machines produce parts of a product that will be assembled once all the K components are available. Let us assume that the time required to produce one component from a machine is modelled by an independent and exponentially distributed random variable with rate μ. The components wait for being assembled in K *join queues*. This is usually reffered as a kitting process. In [17,20] the authors proved that join queue lengths tend to grow infinitely due to the variance of the component production time. Moreover, assuming that the assembly operation is instantaneous, the underlying CTMC $X(t)$ can be studied by means of a dynamically reversible process. It is sufficient to encode the state as a vector $\mathbf{n} = (n_1, \dots, n_K)$ of integer components that represent the difference in the number of pieces produced by the k-th machine and its neighbour k^+ where

$$k^+ = \begin{cases} k+1 & \text{if } k < K \\ 1 & \text{if } k = K. \end{cases}$$

The state space of the model is $\mathcal{S} = \{\mathbf{n} : \sum_{k=1}^{K} n_k = 0 \wedge n_k \in \mathbb{Z}\}$. The expected queue length becomes finite [20] if we can modulate the rates of the component production machines as follows:

$$\mu(n_k) = \begin{cases} \frac{\mu}{n_k+1} & \text{if } n_k \geq 0 \\ \mu & \text{otherwise.} \end{cases}$$

Such a CTMC is dynamically reversible and hence ρ-reversible according to the following renaming function:

$$\rho(\mathbf{n}) = \rho(n_1, \dots, n_K) = (n_K, \dots, n_1) = \mathbf{n}^R,$$

and the equilibrium distribution is given by the expression [20]:

$$\pi(\mathbf{n}) = \frac{1}{G_K} \frac{1}{\prod_{i=1}^{K}(n_i \delta_{n_i > 0})!}, \tag{6}$$

where $\delta_{n_i > 0} = 1$ if n_i is positive, 0 otherwise and G_K is a normalising constant.

Let us assume that we want to change the model such that we impose that the difference between the number of components given by production like k and k^+ is smaller or equal to T, i.e. $n_k \leq T$ for all $k = 1, \dots, K$. The machine that saturates its join queue according to this condition is stopped and will restart working when its neighbour will complete a job. This means that under the immediate assembly time assumption, the maximum join-queue length that we can observe is $(K-1) \cdot T$. Clearly, the CTMC $X'(t)$ underlying such a model is a truncation of the original one, where $\mathcal{A} = \{\mathbf{n} \in \mathcal{S}, n_k \leq T \text{ for all } k = 1, \dots, K\}$. To prove that $X'(t)$ is still ρ-reversible, we use Corollary 4 and we have to show that:

- The partition respects ρ,
- The residence time of $\mathbf{n} \in \mathcal{A}$ and $\mathbf{n}^R$ have the same mean in $X'(t)$.

The first point is easy to prove since if $\mathbf{n} \in \mathcal{A}$ then also $\mathbf{n}^R \in \mathcal{A}$ and vice versa. The second one is trivial since the sum of the arrival rates of the components in $\mathbf{n}$ and $\mathbf{n}^R$ are the same. Therefore, Eq. (6) is an invariant measure for $X'(t)$.

5   Conclusion

In this paper we have studied the aggregation and truncation properties of Markov chains which are reversible modulo a renaming of states. In physics (see e.g., [9–11]) and computer science (e.g., [20, 22]) we can find numerous applications of this theory in the formulation known as *dynamic reversibility*. By the notion of ρ-reversibility, we generalised this definition to arbitrary state renaming functions and showed that the extension is non-trivial, i.e., there are Markov chains which are not dynamically reversible but are ρ-reversible for some ρ which is not an involution. In this paper we have established an important link between ρ-reversibility and the well-known notion of Kelly's reversibility by showing that a certain aggregation of a ρ-reversible chain originates a reversible chain. Although this aggregation is *not* a strong lumping in the sense of Kemeny and Snell work [15], we still have that the aggregated process can be constructed without the computation of the equilibrium distribution of the original chain. Finally, we have revised the well-know results about the truncation of reversible processes in the context of ρ-reversibility and have shown some results that generalise them. Specifically, while the truncation of a reversible chain is always reversible (provided that the irreducibility of the transition graph is maintained) we need some further conditions in order to prove that the truncation of a ρ-reversible chain is also ρ-reversible. These conditions are always trivially satisfied for reversible chains.

A   Proofs of the Results

Proof of Theorem 1

Proof. By Proposition 1 and Definition 1, to prove that $\widetilde{X}(t)$ is $\widetilde{\rho}$-reversible it is sufficient to show that for all $S_i, S_j \in \mathcal{S}/\sim$ with $i \neq j$,

$$\widetilde{\pi}(S_i)\widetilde{q}(S_i, S_j) = \widetilde{\pi}(\widetilde{\rho}(S_j))\widetilde{q}(\widetilde{\rho}(S_j), \widetilde{\rho}(S_i)).$$

By Eq. (3) and Proposition 4, this is equivalent to:

$$\left(\sum_{s \in S_i} \pi(s)\right) \frac{\sum_{s \in S_i} \pi(s) \sum_{s' \in S_j} q(s, s')}{\sum_{s \in S_i} \pi(s)} =$$
$$\left(\sum_{s' \in \widetilde{\rho}(S_j)} \pi(s')\right) \frac{\sum_{s' \in \widetilde{\rho}(S_j)} \pi(s') \sum_{s \in \widetilde{\rho}(S_i)} q(s', s)}{\sum_{s' \in \widetilde{\rho}(S_j)} \pi(s')},$$

which can be written as:

$$\sum_{s \in S_i} \sum_{s' \in S_j} \pi(s) q(s, s') = \sum_{s' \in \widetilde{\rho}(S_j)} \sum_{s \in \widetilde{\rho}(S_i)} \pi(s') q(s', s). \tag{7}$$

We now proceed by considering four cases.

1. Assume that $S_i = \{s\}$ and $S_j = \{s'\}$, then Eq. (7) becomes:

$$\pi(s)q(s, s') = \pi(\rho(s'))q(\rho(s'), \rho(s)),$$

where we have used the definition of $\widetilde{\rho}$ for singletons. This is true since by hypothesis $X(t)$ is ρ-reversible and hence satisfies the ρ-detailed balance equation.

2. Assume $S_i = \{s\}$ and $|S_j| > 1$, and recall that $\widetilde{\rho}(S_j) = S_j$ by definition. Then Eq. (7) can be rewritten as:

$$\sum_{s' \in S_j} \pi(s)q(s, s') = \sum_{s' \in S_j} \pi(s')q(s', \rho(s)).$$

Since $\sim$ respects ρ we have that ρ restricted to the elements of S_j is still a bijection and hence we can write:

$$\sum_{s' \in S_j} \pi(s)q(s, s') = \sum_{s' \in S_j} \pi(\rho(s'))q(\rho(s'), \rho(s)),$$

which is true by the hypothesis of ρ-reversibility of $X(t)$.

3. Assume $|S_i| > 1$ and hence $\widetilde{\rho}(S_i) = S_i$ and $S_j = \{s'\}$, then Eq. (7) can be written as:

$$\sum_{s \in S_i} \pi(s)q(s, s') = \sum_{s \in S_i} \pi(\rho(s'))q(\rho(s'), s).$$

Since ρ restricted to the elements of S_i is a bijection, then we have:

$$\sum_{s \in S_i} \pi(s)q(s, s') = \sum_{s \in S_i} \pi(\rho(s'))q(\rho(s'), \rho(s)),$$

which is an identity.

4. Assume $|S_i| > 1$ and $|S_j| > 1$, and hence $\widetilde{\rho}(S_i) = S_i$ and $\widetilde{\rho}(S_j) = S_j$. Then we can rewrite Eq. (7) as:

$$\sum_{s \in S_i} \sum_{s' \in S_j} \pi(s)q(s, s') = \sum_{s \in S_i} \sum_{s' \in S_j} \pi(s')q(s', s).$$

Since ρ restricted to S_i and to S_j is still a bijection because $\sim$ respects ρ, we can rewrite the previous equation as:

$$\sum_{s \in S_i} \sum_{s' \in S_j} \pi(s)q(s, s') = \sum_{s \in S_i} \sum_{s' \in S_j} \pi(\rho(s'))q(\rho(s'), \rho(s)).$$

which is true by hypothesis. □

Proof or Proposition 5

Proof. By the general aggregation Eq. (3), for any $S_i, S_j \in \mathcal{S}/\sim$,

$$\widetilde{q}(S_i, S_j) = \frac{\sum_{s' \in S_i} \pi(s') \sum_{s \in S_j} q(s', s)}{\sum_{s' \in S_i} \pi(s')}, \tag{8}$$

Since $X(t)$ is ρ-reversible and each $S_i \in \mathcal{S}/\sim$ is an orbit for ρ, it holds that $\pi(s) = \pi(s')$ for all $s, s' \in S_i$. Let us denote by $\pi(S_i)$ the equilibrium probability of each s belonging to the orbit S_i. Hence, $\sum_{s' \in S_i} \pi(s') = |S_i| \pi(S_i)$ and Eq. (8) can be written

$$\widetilde{q}(S_i, S_j) = \pi(S_i) \frac{\sum_{s' \in S_i} \sum_{s \in S_j} q(s', s)}{|S_i| \pi(S_i)} \tag{9}$$

proving the statement. □

Proof of Lemma 1

Proof. To prove the lemma we use Proposition 1. In fact, let us consider two states $s, u \in \mathcal{A}$, then the corresponding ρ-detailed balance equation is $B\pi(s)q(s, u) = B\pi(\rho(u))q(\rho(u), \rho(s))$ since we have by assumption that the partition respects ρ and hence also $\rho(t), \rho(s) \in \mathcal{A}$. This equation is satisfied because $X(t)$ is ρ-reversible. If $s, u \in \mathcal{S} \setminus \mathcal{A}$ the corresponding detailed balance equation is $Bc\pi(s)q(s, u) = Bc\pi(\rho(u))q(\rho(u), \rho(s))$ that is also satisfied for the same reasons. Let us consider $s \in \mathcal{A}$ and $u \in \mathcal{S} \setminus \mathcal{A}$, then we have that the transition rates are modified and hence $B\pi(s)(cq(s, u)) = Bc\pi(\rho(u))q(\rho(u), \rho(s))$ which is an identity since $\sim$ respects ρ. Finally, we have to consider the case of $s \in \mathcal{S} \setminus \mathcal{A}$ and $u \in \mathcal{A}$. The corresponding detailed balance equation is $Bc\pi(s)q(s, u) = B\pi(\rho(u))(cq(\rho(u), \rho(s)))$ which is satisfied by hypothesis. The fact that the residence times in the states belonging to the same orbits of ρ in $X'(t)$ are identically distributed is an assumption of the lemma. □

References

1. Akyildiz, I.F.: Exact analysis of queueing networks with rejection blocking. In: Perros, H.G., Atliok, T. (eds.) Proceedings of the 1st International Workshop on Queueing Networks with Blocking, pp. 19–29 (1989)
2. Balsamo, S., Marin, A.: Separable solutions for Markov processes in random environments. Eur. J. Oper. Res. **229**(2), 391–403 (2013)
3. Balsamo, S., Dei Rossi, G., Marin, A.: Lumping and reversed processes in cooperating automata. Ann. Oper. Res. **239**, 695–722 (2014)
4. Benaim, M., Le Boudec, J.-Y.: A class of mean field interaction models for computer and communication systems. Perform. Eval. **65**(11–12), 823–838 (2008)
5. Block, R., Van Houdt, B.: Spatial fairness in multi-channel CSMA line networks. In: Proceedings of the 8th International Conference on Performance Evaluation Methodologies and Tools, VALUETOOLS, pp. 1–8 (2014)
6. Buchholz, P.: Product form approximations for communicating Markov processes. Perform. Eval. **67**(9), 797–815 (2010). Special Issue: QEST 2008

7. Bujari, A., Marin, A., Palazzi, C.E., Rossi, S.: Analysis of ECN/RED and SAP-LAW with simultaneous TCP and UDP traffic. Comput. Netw. **108**, 160–170 (2016)
8. Comert, G.: Queue length estimation from probe vehicles at isolated intersections: estimators for primary parameters. Eur. J. Oper. Res. **252**(2), 502–521 (2016)
9. Gates, D.J.: Growth and decrescence of two-dimensional crystals: a Markov rate process. J. Stat. Phys. **52**(1/2), 245–257 (1988)
10. Gates, D.J., Westcott, M.: Kinetics of polymer crystallization I. Discrete and continuum models. Proc. R. Soc. Lond. **416**, 443–461 (1988)
11. Gates, D.J., Westcott, M.: Markovian models of steady crystal growth. J. Appl. Prob. **3**(2), 339–355 (1993)
12. Gelenbe, E., Labed, A.: G-Networks with multiple classes of signals and positive customers. Eur. J. Oper. Res. **48**(5), 293–305 (1998)
13. Kelly, F.: Reversibility and Stochastic Networks. Wiley, New York (1979)
14. Kelly, F.: Loss networks. Ann. Appl. Probab. **1**(3), 319–378 (1991)
15. Kemeny, J.G., Snell, J.L.: Finite Markov Chains. Springer, Heidelberg (1976)
16. Kozlowski, D., Worthington, D.: Use of queue modelling in the analysis of elective patient treatment governed by a maximum waiting time policy. Eur. J. Oper. Res. **244**(1), 331–338 (2015)
17. Latouche, G.: Queues with paired customers. J. Appl. Probab. **18**(3), 684–696 (1981)
18. Marin, A., Rossi, S.: On discrete time reversibility modulo state renaming and its applications. In: Proceedings of the 8th International Conference on Performance Evaluation Methodologies and Tools, VALUETOOLS, pp. 1–8 (2014)
19. Marin, A., Rossi, S.: On the relations between lumpability and reversibility. In: Proceedings of the IEEE 22nd International Symposium on Modeling, Analysis and Simulation of Computer and Telecommunication Systems (MASCOTS 2014), pp. 427–432 (2014)
20. Marin, A., Rossi, S.: Dynamic control of the join-queue lengths in saturated fork-join stations. In: Agha, G., Houdt, B. (eds.) QEST 2016. LNCS, vol. 9826, pp. 123–138. Springer, Cham (2016). doi:10.1007/978-3-319-43425-4_8
21. Marin, A., Rossi, S.: On the relations between Markov chain lumpability and reversibility. Acta Inform., 1–39 (2016). doi:10.1007/s00236-016-0266-1
22. Pan, R., Prabhakar, B., Psounis, K.: CHOKe, a stateless active queue management scheme for approximating fair bandwidth allocation. In: Proceedings of IEEE INFOCOM 2000, pp. 942–951. IEEE Computer Society Press, Washington, DC (2000)
23. Roy, D., Krishnamurthy, A., Heragu, S.S., Malmborg, C.J.: Queuing models to analyze dwell-point and cross-aisle location in autonomous vehicle-based warehouse systems. Eur. J. Oper. Res. **242**(1), 72–87 (2015)
24. Shone, R., Knight, V.A., Williams, J.E.: Comparisons between observable and unobservable M/M/1 queues with respect to optimal customer behavior. Eur. J. Oper. Res. **227**(1), 133–141 (2013)
25. Whittle, P.: Systems in Stochastic Equilibrium. Wiley, New York (1986)

Modeling Multiclass Task-Based Applications on Heterogeneous Distributed Environments

Riccardo Pinciroli[✉], Marco Gribaudo, and Giuseppe Serazzi

Dipartimento di Elettronica, Informazione e Bioingengeria, Politecnico di Milano,
via Ponzio 34/5, 20133 Milano, Italy
{riccardo.pinciroli,marco.gribaudo,giuseppe.serazzi}@polimi.it

Abstract. The volume of data, one of the five "V" characteristics of Big Data, grows at a rate that is much higher than the increase of ability of the existing systems to manage it within an acceptable time. Several technologies have been developed to approach this scalability issue. For instance, MapReduce has been introduced to cope with the problem of processing a huge amount of data, by splitting the computation into a set of tasks that are concurrently executed. The savings of even a marginal time in the processing of all the tasks of a set can bring valuable benefits to the execution of the whole application and to the management costs of the entire data center. To this end, we propose a technique to minimize the global processing time of a set of tasks, having different service requirements, concurrently executed on two or more heterogeneous systems. The validity of the proposed technique is demonstrated using a multiformalism model that consists of a combination of Queueing Networks and Petri Nets. Application of this technique to an Apache Hive case-study shows that the described allocation policy can lead to performance gains on both total execution time and energy consumption.

Keywords: Pool depletion systems · MapReduce · Schedulers · Energy efficiency · Performance evaluation · Queueing networks · Petri nets · Multiformalism models

1 Introduction

The pervasiveness of Big Data applications in organizations is occurring at a surprising high speed. Successful companies must adopt these new technologies in order to keep their advantage over competitors.

One of the most important characteristics of this new paradigm is the large size of data that must be processed in a reasonable amount of time. To address the resulting performance problem that is created, the Hadoop MapReduce technology has been proposed [9,14] originally by Google. Its operational concept is based on distributed computing and parallelism. Initially, the input data is split into blocks that are processed in parallel by a high number of tasks generated by each job in the Map. In the following Reduce phase, newly created tasks process

© Springer International Publishing AG 2017
N. Thomas and M. Forshaw (Eds.): ASMTA 2017, LNCS 10378, pp. 166–180, 2017.
DOI: 10.1007/978-3-319-61428-1_12

in parallel the intermediate results of the Map phase producing the final output of the job. The execution of a job may require one or more cycles of Map and Reduce phases. Typically, each phase can take hours, or even days, to complete due to the significantly large data sizes and the consequent high number of tasks to be executed in parallel.

In [20] it is described an example of an Apache Hive application (see Fig. 5) regarding a query that retrieves the Facebook status according to users' gender and school. Its structure is based on three MapReduce jobs. The two Maps in *MapReduce1* generate a high number of tasks that, as a function of the query, have different resource consumption, i.e., they belong to two classes of tasks. The tasks are then assigned to Virtual Machines and executed in parallel on the same, or more likely, on different physical systems.

The analysis of resource consumption behavior during the execution of a MapReduce job shows a pattern that is repeated in each phase: generation of a large number of tasks, followed by their parallel execution. Typically, the number of parallel tasks is much higher in the Map phase with respect to the Reduce one.

The execution time of the parallel tasks is deeply influenced by two factors that are related to the characteristics of both the job and the architecture of the computing infrastructure. The first factor concerns the resource requirements of the tasks, since each phase of a job may saturate different resources, i.e., it can have a different bottleneck. The second factor regards the characteristics of the physical systems that are used. Indeed, in cloud infrastructures, the computers that are dynamically allocated to the tasks of a job may be heterogeneous and have different computational power and storage capacity (see e.g., [6]). Furthermore, these physical machines typically have a limitation on the number of tasks to be executed. This constraint, continuously reached in MapReduce jobs due to the high number of tasks, is required for performance control on response times. As a function of the mix of tasks in a single computer (referred to as subsystem), the time required by their parallel execution may be in some cases extremely inflated due to the bottlenecks that may migrate dynamically among the resources. The variability introduced in the execution time of the Map tasks may have a large impact on the execution time of a job. To cope with this problem, the tasks admission policy in each subsystem plays a fundamental role.

In order to analyze the performance of Big Data applications we study the *Pool depletion systems* [7,8]. A Pool depletion system is a framework composed by a pool of tasks and by several subsystems. We assume that the tasks are created in the pool at the same instant of time and then are sent to several parallel subsystems for their execution. The important parameter is the time required by the execution of all the tasks, referred to as depletion time. We may consider a pool as a container for the Map tasks, all of which must be completely executed before starting the following Reduce phase. In this paper, we describe a scheduling policy that minimizes the depletion time of a *pool* of tasks by appropriately allocating them on heterogeneous systems with limited capacity as a function of the different resource requirements. In particular, our technique increases the efficiency of the global system allowing the resources to operate simultaneously at their optimal conditions. Several scheduling strategies have been described in literature to deal with systems that serve a large number of tasks. The objectives of the proposed policies are: (i) to minimize the execution time of each task; (ii) to optimize the system utilization exploiting the jobs allocation on each resource, (iii) to minimize the execution time of a single job that is served by all the resources of the system, as may happen in scientific applications. *Case i* has been deeply analyzed in literature. FIFO, JSQ, MaxWeight and Completely Fair Scheduler strategies are some examples of scheduling policies used in that case. Unfortunately, in the problem that we consider, minimizing the task execution time does not necessarily minimize the job execution. Also *case ii* has been studied in depth. For example, Fair and Capacity scheduler algorithms [18] may be adopted by Hadoop to allocate resources in order to improve system utilization when executing jobs possibly from multiple tenants. The objective of our problem is described by *case iii*. The *Optimal population mix* [19] policy introduced in [8].

In our case, each subsystem is composed by two resources, e.g., CPU and storage. We consider a multi-class workload, with two classes of jobs. From [19] we know that, for a system with two resources and two classes of customers, a mix of classes in execution that maximizes the utilization of all the resources exists, and it is referred to as *optimal population mix*. Furthermore, we can analytically derive the *optimal population mix* for one-subsystem Pool depletion systems [8]. When the system operates with that mix, the pool can be depleted with the shortest time. Two main configurations for multi-subsystem Pool depletion systems are analyzed: the homogeneous and the heterogeneous subsystems configurations. In the former case, all the available subsystems are identical, and their corresponding resources have the same characteristics (i.e., service demands). In the latter case, each subsystem may be different from all the others, thus the resources have different service demands. In particular, the heterogeneous configuration may be used for modeling a cluster with servers having different capacity. To compute the results, we model the applications with a multiformalism model composed by queuing network and Petri nets. We simulated a variable number of subsystems and different internal population mixes, while the pool population mix is set to a constant value corresponding to the considered application.

The remainder of this paper is structured as follows. In Sect. 2 prior work on Pool depletion systems is discussed. Section 3 extends Pool depletion systems with multi-subsystem and presents the multiformalism model used for the extension. The results for homogeneous and heterogeneous multi-subsystem are shown in Sect. 4. An exploitation of this kind of framework is shown in Sect. 5 for the analysis of an Apache Hive case-study. Section 6 concludes the paper.

2   Background

The main applications we want to study through Pool depletion systems are the ones used in Big Data environments. Although the key feature of these applications is the parallel execution of a high number of processes, for sake of simplicity the Pool depletion systems have been analyzed so far [7,8] with only one-subsystem configuration. In [7] Pool depletion systems are described, and a Continuous time Markov chain (CTMC) is proposed to investigate multi-class cases. In particular, trade-off between energy consumption and system performance has been evaluated using some of the most common metrics, such as: the product of energy consumption and response time, or Energy-Response time product (ERP) [11,16,17]; their sum, or Energy-Response time weighted sum (ERWS) [1,2,15]; the average energy consumed per job, or Energy per job (EJ) [5,13]. Two-class Pool depletion systems are analytically investigated in [8], and closed-form equations for the derivation of the optimal population mixes of pool and subsystem are provided. For this purpose, the definitions [19] of *equi-utilization point* (i.e., the population mix of a system for which all its resources are equi-utilized) and *equi-load point* (i.e., the population mix for which all the resources of a system are equally loaded) are used. In particular, the optimal population mix of a subsystem, β^*, corresponds to the *equi-utilization point* of the subsystem itself, and for a subsystem with two resources and two classes it is derived as follows:

$$\beta^* = \left(\beta_A^* = \frac{\log \frac{D_{2B}}{D_{1B}}}{\log \frac{D_{1A}D_{2B}}{D_{1B}D_{2A}}}, \ \beta_B^* = 1 - \beta_A^* \right) \tag{1}$$

where D_{rc} is the service demand of a class c job at resource r. The pool optimal population mix, $\boldsymbol{\alpha}^*$, corresponds to the *equi-load point* of a closed system, and may be computed as:

$$\boldsymbol{\alpha}^* = \left(\alpha_A^* = \frac{X_A(\boldsymbol{\beta}^*)}{X_A(\boldsymbol{\beta}^*) + X_B(\boldsymbol{\beta}^*)}, \ \alpha_B^* = 1 - \alpha_A^* \right) \tag{2}$$

where, $X_c(\boldsymbol{\beta}^*)$ is the throughput of class c jobs when the subsystem population mix is the optimal one. It has also been proved that depletion time (i.e., the time needed to completely execute all the requests initially in the pool) is minimized when the Pool depletion system works with its optimal population mixes. Note that, while subsystem population mix is a feature of the system and β^* may be exploited by a smart scheduler to make the whole system work with better performance, pool population mix depends on the type of application we are considering and can change only if the application changes.

In this paper we focus on multi-subsystem Pool depletion systems, where the requests in the pool are executed by several parallel subsystems. The considered subsystems are heterogeneous and have constraints on the number of tasks executed simultaneously. A similar approach was adopted in [5]. In that case, differently from this paper, a multi-class *open network* was taken into consideration. A smart scheduler was implemented to decrease the energy consumption of the global system. In particular, it forwarded each incoming job to one of its subsystems, depending on their current population mix.

3 The Model

The considered Pool depletion system is described using the model consisting of a Coloured Petri Net (CPN) and a multi-class fork-and-join queuing network shown in Fig. 1. The workload of the model consists of two type of customers: the tokens, representing the colours of the Petri nets, and the jobs, representing the requests to be executed by the queueing networks. Each type of workload comprises two different classes of customers. CPN tokens are used to control the access to resources by the tasks, while queuing networks primitives are used to describe the service demands of the tasks. According to [12], the use of several formalisms allows exploiting the most appropriate modelling primitives to express the corresponding concepts in the most efficient and natural way.

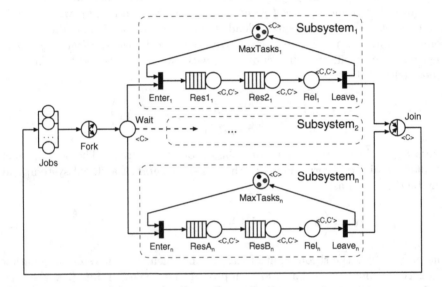

Fig. 1. The multiformailsm model of the considered scenario

The jobs are created in the delay station Jobs: in this work we will focus, without loss of generality, on a single job that is continuously executed, and we set the "think time" (service time of the Jobs station) to $Z = 0$. The job enters the Fork node and it is splitted into N_A tasks of class A and N_B tasks of class B, with $N = N_A + N_B$. Tasks waiting to be executed are represented as tokens into place Wait: in this case, colour class $<C>$ is used to remember the task class as an attribute of the token. Table 1 summarises the colour classes used in the model; in Fig. 1 colour classes are represented, in angled brackets, as labels associated to places and, with a slight abuse of notation, to queuing stations.

The place MaxTasks represents the considered scheduling assignment and contains tokens belonging to the $<C>$ colour class, initialised to K_A and K_B tokens for the corresponding task classes. Its marking represents the chosen

Table 1. Colour-sets.

Colour-set	Description
$<C>$	Task class $C = \{A, B\}$.
$<C, C'>$	Task class C, original class C'/

configuration that allows a total of $K = K_A + K_B$ tasks simultaneously in a subsystem. The execution of a task starts with a firing of transition `Enter`, which can occur in one of the four following modes. When the subsystem is in normal operation, transition can fire in mode 1 or 2 removing respectively one token of class A or B from its input place, and creating customers of class (A, A) or (B, B) in the queue `Res1`, that represents the first resource of the subsystem. When there are no more tokens of either class B or A in place `Wait`, transition `Enter` fires in mode 3 or 4 allowing a task of class A or B to enter instead of the one that has already been completed. Queuing stations `Res1` and `Res2` represent the two resources of the subsystem. Even if the task classes generated by the CPN transition are $(A, A), (A, B), (B, B)$ and (B, A), only the first component of each couple is used to determine the service requirements of one task. The second component is instead used in place `Rel` to allow transition `Leave` forwarding the correct task type to the `Join` node, and to return the acquired task token into place `MaxTasks`. This is accomplished by the firing of transition `Leave` according to four modes, as summarised in Table 2.

Table 2. Transitions firing modes.

Transition	Mode	In$_1$	In$_2$	Out$_1$	Out$_2$	Description
`Enter`		`Wait`	`MaxTasks`	`Res1`		
	1	A	A	(A, A)		Class A task
	2	B	B	(B, B)		Class B task
	3	A	B with `Wait`.$B = 0$	(A, B)		Class A task, depletion
	4	B	A with `Wait`.$A = 0$	(B, A)		Class B task, depletion
`Leave`		`Rel`		`MaxTasks`	`Join`	
	1	(A, A)		A	A	Class A task
	2	(B, B)		B	B	Class B task
	3	(A, B)		B	A	Class A task, depletion
	4	(B, A)		A	B	Class B task, depletion

Each subsystem has a similar structure, and the same sub-model is repeated n times as shown in Fig. 1. Each subsystem can be characterised by different service demand for its resources, and different tasks allowances K_A and K_B. When all tasks have been completed, the `Join` primitive can fire, returning the customer to the reference station, and allowing the next job to start. To characterise the configuration of the system, we will denote with:

$$\alpha = \left(\alpha_A = \frac{N_A}{N_A + N_B}, \ \alpha_B = 1 - \alpha_A \right)$$

$$\beta = \left(\beta_A = \frac{K_A}{K_A + K_B}, \ \beta_B = 1 - \beta_A \right)$$

the *pool population mix* and the *subsystem population mix*, respectively, as described in Eqs. (1) and (2).

4   Results

In this section, the results obtained analyzing multi-class applications on distributed environments are shown. In [8], we showed that analytical equations – to identify the optimal point for a single-machine system – exist. Eqs. (1) and (2) show the results for subsystem and pool of a Pool depletion system, respectively. While considering distributed environments does not affect Eq. (1) since it refers to each single subsystem, it influences the optimal population mix of the pool in Eq. (2). In fact, as said in Sect. 2, optimal population mix of the pool is related to the equi-load point of the system and it varies when we take into account several different subsystems. In particular, we expand α_A^* as follows:

$$\alpha_A^* = \frac{X_A(\beta^*)}{X_A(\beta^*) + X_B(\beta^*)} = \frac{\sum_{s=1}^n X_A^s(\beta^{s*})}{\sum_{s=1}^n X_A^s(\beta^{s*}) + \sum_{s=1}^n X_B^s(\beta^{s*})} \tag{3}$$

where n is the total number of subsystems, and $X_c^s(\beta^{s*})$ is the throughput for class c jobs at subsystem s, when the population mix of that subsystem is its optimal population mix, β^{s*}.

Two main system configurations are considered. First of all, the effects of homogeneous distributed systems – where all the subsystems are identical – are taken into account. Then, we consider heterogeneous distributed environments, assuming that each subsystem may have different characteristics with respect to the other ones. The analyses are performed using JSIMgraph, the JMT [4] simulation tool, and simulating the model described in Sect. 3. All the metrics are computed with 99% confidence interval and a 3% maximum relative error.

4.1   Homogeneous

The main advantage introduced by homogeneous distributed systems is the parallelization of jobs execution. Since all the subsystems are the same, their optimal population mix, β^{s*}, are identical. Due to that, the throughput of each subsystem does not change and Eq. (3), after some algebraic manipulation, becomes equal to Eq. (2).

In Fig. 2 three performance metrics for a homogeneous distributed system are depicted against the number n of subsystems, and their population mix, β. The pool size of the considered system is $N = 1008$ and the capacity of each subsystem is $K = 108/n$. In the simulations we consider $n = \{1, 2, 3, 4, 6\}$

and the following service demand matrix for each two-resource and two-class subsystem:

$$D_{rc} = \begin{bmatrix} 0.75 & 0.64 \\ 0.48 & 1.25 \end{bmatrix} \qquad (4)$$

The optimal population mix of these subsystems is computed through Eq. (1) and it is $\boldsymbol{\beta}^* = (0.6, 0.4)$. Their *equi-load point* is $\boldsymbol{\alpha}^* = (0.693, 0.307)$. Since all the subsystems are identical, we assume they are all set to work with the same population mix. For sake of simplicity, we only consider an application with $N_A = 724, N_B = 284$ and $\boldsymbol{\alpha} = \left(\frac{724}{1008}, \frac{284}{1008}\right)$.

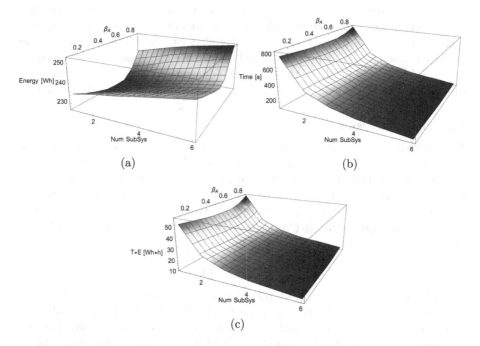

Fig. 2. Energy consumption (a), depletion time (b) and Time · Energy (c) of a homogeneous distributed system for $\boldsymbol{\alpha} = \left(\frac{724}{1008}, \frac{284}{1008}\right)$, as a function of total number of subsystems and their population mix.

Figure 2a represents the energy consumed to serve all the tasks initially in the pool when $P_{idle}^1 = 315\,\text{W}, P_{busy}^1 = 630\,\text{W}, P_{idle}^2 = 250\,\text{W}$ and $P_{busy}^2 = 500\,\text{W}$. P_{idle}^r and P_{busy}^r are the power consumption of a resource r when it is idle and fully utilized, respectively. The energy consumed by a Pool depletion system is defined as the product between depletion time and power, $E = T \cdot P$, and the power consumption is computed with equation

$$P(U) = P_{idle} + (P_{busy} - P_{idle}) \cdot U \qquad (5)$$

proposed by Fan et al. in [10]. In our case, the minimum value of energy consumption is observed when the system is working with only one subsystem. However, it is interesting to note that working with six subsystems and with the optimal population mix, β^*, lets the service provider save more energy than working with only one subsystem and with a suboptimal population mix. Figure 2b shows the depletion time of the system (i.e., the time needed by the system to complete all the tasks initially in the pool). In this case, providing a large number of subsystems allows the service provider to parallelize the work, and with six subsystems the depletion time is four times lower than with only one subsystem. In order to take into consideration both the measures (i.e., energy consumption and depletion time), we compute their product and the results are shown in Fig. 2c. Since the saved amount of time working with six subsystems is definitely greater than the energy the system consumes working with only one machine, the configuration with the larger amount of subsystems is the one with the best global performance, assuming that energy is as important as the depletion time for service provider. Furthermore, for the considered application, i.e., $\alpha = \left(\frac{724}{1008}, \frac{284}{1008} \right)$, all the three analyzed metrics have their minimum point when all the available subsystems work with their optimal population mix, β^*.

4.2   Heterogeneous

The analysis of heterogeneous distributed environements is interesting since it lets us consider more general systems. For example, they can be used to model a datacenter with different types of servers (e.g., new vs. old, fast vs. slow, etc.). In order to study this kind of environments, we focus on the two-subsystem case. Since they must have different features, the first subsystem is defined through service demand matrix in Eq. (4), whereas the following matrix is considered for the second subsystem:

$$D_{rc} = \begin{bmatrix} 0.86 & 0.65 \\ 0.3 & 1.02 \end{bmatrix} \tag{6}$$

The optimal population mix of the subsystem defined in Eq. (6) is $\beta^{2*} = (0.3, \; 0.7)$ and its *equi-load point* is $\alpha^{2*} = (0.409, \; 0.693)$. They have been derived through Eqs. (1) and (2), respectively. As previously said, the optimal population mix of the pool must be computed with Eq. (3) if the system has heterogeneous subsystems, and in this case it is $\alpha^* \simeq (0.55, \; 0.45)$. This result highlights another important gain of a heterogeneous distributed environment, besides the parallelization of the workload: indeed, differently from the homogeneous case, the implementation of inhomogeneous subsystems lets the service provider execute different applications with better performance. This is shown in Fig. 3 that depicts the depletion times of two different applications (i.e., pool population mixes) served by a homogeneous system and a heterogeneous one. The population mixes of the two subsystems is still assumed to be the same for both of them. As said, it is interesting to note that the heterogeneous configuration provides a shorter depletion time than the homogeneous one when it is serving application $\alpha = (0.55, \; 0.45) \simeq \alpha^*_{heter}$. On the contrary, if application $\alpha = (0.70, \; 0.30)$ is considered, the homogeneous environment performs better than the heterogeneous one, since the new application is closer to the homogeneous optimal pool population mix, α^*_{homo}.

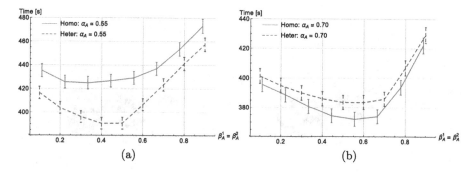

Fig. 3. Comparison of depletion time of homogenenous and heterogeneous systems for two different applications: $\alpha = (0.55,\ 0.45)$ (a) and $\alpha = (0.70,\ 0.30)$ (b).

Energy consumption, depletion time and their product are depicted in Fig. 4. Each metric is analyzed against the population mix of each subsystem (i.e., β^1 and β^2). For all the three metrics, the minimum values are registered when each subsystem s works with its optimal population mix, β^{s*}. Also in this case, the amount of time saved to complete all the tasks initially in the pool is definitely larger than the energy saved, thus depletion time affects the system performance more than energy consumption when both the metrics have the same weight on the service provider's decisions.

5   Exploitation: Apache Hive

In this section we apply the results obtained in Sect. 4 to a real Big Data application. Simulations have been performed for both FIFO and *Optimal population mix* admission policies. Although Pool depletion systems can model many different Big Data applications, the case-study we consider in this paper is an Apache Hive [20] based application. Apache Hive is an open source data warehouse system and is used on top of Apache Hadoop. It has been extensively adopted by many organizations (e.g., Facebook, Netflix, Spotify) since it makes easier the management of large data and their queries. Hive was introduced by Facebook in [20]. It can be run on different Hadoop's framework (e.g., MapReduce, YARN, Spark, Tez) and provides a SQL-like query language that is called HiveQL. A query to retrieve most popular Facebook status based on users' gender and school is used as an example in [20]. Its simplified query plan is shown in Fig. 5, where three MapReduce jobs are represented.

In order to study the performance of the *Optimal population mix* strategy shown in Sect. 4, we compare the time this strategy needs to complete a MapReduce job with the time required by the FIFO strategy for the same kind of job. We focus on the join function of a Hive query, that is represented by *MapReduce1* in Fig. 5. In that case, both Map and Reduce have multi-class workloads. Indeed, Map phase gets its data from two different tables, whereas the Reduce one returns two temporary tables. Assuming that two tables must be joint, we

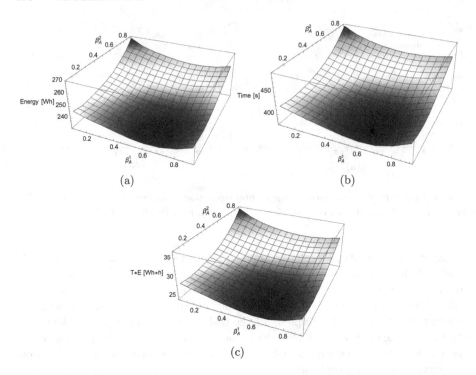

Fig. 4. Energy consumption (a), depletion time (b) and Time · Energy (c) of a heterogeneous distributed system for $\alpha = (0.55,\ 0.45)$, as a function of population mix of each subsystem.

want to identify the best way to execute all the tasks of a MapReduce job in order to decrease the total time needed to complete the join clause and get the expected results. The Map and Reduce phases, are modeled by two different heterogeneous two-subsystem environments, that have been characterized starting from [21] and [3]. The former provides some data about MapReduce jobs executed on Facebook's clusters, whereas the latter refers to a public Hadoop repository[1]. Thus, the Map system is defined by the following service demands matrices (in seconds):

$$D_{rc}^{1,Map} = \begin{bmatrix} 25 & 21 \\ 16 & 39 \end{bmatrix} \qquad D_{rc}^{2,Map} = \begin{bmatrix} 55 & 39 \\ 28 & 82 \end{bmatrix} \tag{7}$$

whereas the Reduce one has the following service demands (still in seconds):

$$D_{rc}^{1,Reduce} = \begin{bmatrix} 60 & 46 \\ 17 & 70 \end{bmatrix} \qquad D_{rc}^{1,Reduce} = \begin{bmatrix} 29 & 22 \\ 10 & 34 \end{bmatrix} \tag{8}$$

We assume the MapReduce job is splitted into $N_{Map} = 1000$ and $N_{Red} = 200$ tasks before Map and Reduce phases, respectively. The number of tasks that is

---

[1] Available at http://ftp.pdl.cmu.edu/pub/datasets/hla/. Please, include *http* at the beginning of the URL to make it work.

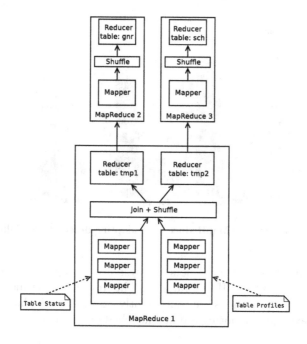

Fig. 5. Query plan of three MapReduce jobs analyzed in [20].

concurrently executed during the two phases is $K_{Map} = 100$ for Map and $K_{Red} = 50$ for Reduce. Using Eq. (1) on all the service demand matrices, we derive the optimal population mixes of each subsystem, i.e., $\beta^{1*}_{Map} = (0.58, 0.42)$, $\beta^{2*}_{Map} = (0.29, 0.71)$, $\beta^{1*}_{Red} = (0.52, 0.48)$ and $\beta^{2*}_{Red} = (0.26, 0.74)$. Finally, we take into account an application whose $\alpha_{Map} = (0.53, 0.47)$ for Map phase and $\alpha_{Red} = (0.49, 0.51)$ for the Reduce one. For both Map and Reduce, $\alpha \simeq \alpha^*$ has been computed through Eq. (3).

The results of the simulations have been computed with 99% confidence interval, and they are shown in Fig. 6. Besides the *Optimal population mix* strategy, we analyze two FIFO policies: we consider completion of all *class A* tasks before starting executing *class B* ones, and viceversa.

In Fig. 6 the time to complete all the Map and Reduce tasks into the system are compared based on the scheduling policy adopted by the scheduler. The total time to complete a MapReduce job is also depicted. Due to the large number of tasks we assume to be into the system during the Map, this phase is affecting the global performance of the system more than the Reduce one. When the subsystems work with their optimal population mixes, the service provider complete all the tasks (i.e., the MapReduce job) in a shorter time than using a FIFO strategy. The system requires the longest time to complete all the tasks when it serves *class B* tasks after completing all the *class A* ones. For sake of simplicity, we assume Reduce tasks are served after the Map ones, and no parallelism between tasks belonging to different phases is admitted (i.e., *Total length = Map phase length + Reduce phase length*). Based on the observed results,

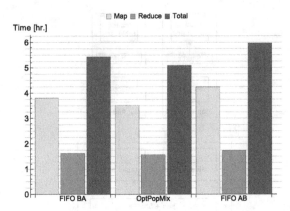

Fig. 6. Time to complete the MapReduce jobs adopting different scheduling strategies.

Optimal population mix strategy lets the system save up to 15% on the total time, with respect to the FIFO strategies. This result is even more interesting when several MapReduce jobs are concurrently executed and a larger amount of time can be saved just adopting a different scheduling policy.

6   Conclusion

In this paper we investigated the performance gain introduced by the adoption of an *Optimal population mix* scheduling strategy in the execution of Big Data applications. In order to do that, Pool depletion systems have been extended with the implementation of two or more subsystems. This extension of Pool depletion systems was necessary since parallel tasks execution is the key feature of the Big Data applications. We identified two main system configurations: the homogeneous and the heterogeneous ones. In the former, all the subsystems are identical and the only advantage with respect to the single-subsystem Pool depletion systems is the possibility to parallelize the execution of the tasks. In the latter case, instead, the subsystems may have different characteristics (i.e., service demands). The service provider may exploit this other feature in order to process some applications (defined by the number of the *Class A* and *Class B* tasks from which they are composed) with better performance.

The analysis of multi-subsystems Pool depletion systems provided in this paper highlights that depletion time (i.e., the time needed to complete all the tasks initially in the pool) affects the global performance of these systems more than the energy consumed to run a larger number of subsystems.

Finally, the *Optimal population mix* scheduling strategy is compared with the default FIFO policy. The comparison is performed considering a specific MapReduce job: Facebook's Apache Hive query. From the simulation of the case-study and the anlysis of its results, the implementation of the *Optimal population mix* policy lets the service provider save up to the 15% on the total amount of time needed to complete the MapReduce job.

Acknowledgement. This research was supported in part by the European Commission under the grant ANTAREX H2020 FET-HPC-671623.

References

1. Andrew, L.L., Lin, M., Wierman, A.: Optimality, fairness, and robustness in speed scaling designs. In: ACM SIGMETRICS Performance Evaluation Review, vol. 38, pp. 37–48. ACM (2010)
2. Bansal, N., Chan, H.L., Pruhs, K.: Speed scaling with an arbitrary power function. In: Proceedings of the Twentieth Annual ACM-SIAM Symposium on Discrete Algorithms, pp. 693–701. Society for Industrial and Applied Mathematics (2009)
3. Barbierato, E., Gribaudo, M., Manini, D.: Fluid approximation of pool depletion systems. In: Wittevrongel, S., Phung-Duc, T. (eds.) ASMTA 2016. LNCS, vol. 9845, pp. 60–75. Springer, Cham (2016). doi:10.1007/978-3-319-43904-4_5
4. Bertoli, M., Casale, G., Serazzi, G.: JMT: performance engineering tools for system modeling. SIGMETRICS Perform. Eval. Rev. **36**(4), 10–15 (2009)
5. Cerotti, D., Gribaudo, M., Piazzolla, P., Pinciroli, R., Serazzi, G.: Multi-class queuing networks models for energy optimization. In: Proceedings of the 8th International Conference on Performance Evaluation Methodologies and Tools, pp. 98–105. ICST (Institute for Computer Sciences, Social-Informatics and Telecommunications Engineering) (2014)
6. Cerotti, D., Gribaudo, M., Piazzolla, P., Serazzi, G.: Flexible CPU provisioning in clouds: a new source of performance unpredictability. In: Ninth International Conference on Quantitative Evaluation of Systems, QEST 2012, London, United Kingdom, 17–20 September 2012, pp. 230–237 (2012)
7. Cerotti, D., Gribaudo, M., Pinciroli, R., Serazzi, G.: Stochastic analysis of energy consumption in pool depletion systems. In: Remke, A., Haverkort, B.R. (eds.) MMB&DFT 2016. LNCS, vol. 9629, pp. 25–39. Springer, Cham (2016). doi:10.1007/978-3-319-31559-1_4
8. Cerotti, D., Gribaudo, M., Pinciroli, R., Serazzi, G.: Optimal population mix in pool depletion systems with two-class workload. In: 10th EAI International Conference on Performance Evaluation Methodologies and Tools. ACM (2017)
9. Dean, J., Ghemawat, S.: MapReduce: simplified data processing on large clusters. Commun. ACM **51**(1), 107–113 (2008)
10. Fan, X., Weber, W.D., Barroso, L.A.: Power provisioning for a warehouse-sized computer. In: ACM SIGARCH Computer Architecture News, vol. 35, pp. 13–23. ACM (2007)
11. Gandhi, A., Gupta, V., Harchol-Balter, M., Kozuch, M.A.: Optimality analysis of energy-performance trade-off for server farm management. Perform. Eval. **67**(11), 1155–1171 (2010)
12. Gribaudo, M., Iacono, M.: Theory and Application of Multi-formalism Modeling. IGI Global, Hershey (2013)
13. Ho, T.T.N., Gribaudo, M., Pernici, B.: Characterizing energy per job in cloud applications. Electronics **5**(4), 90 (2016)
14. Huang, L., Wang, X.W., Zhai, Y.D., Yang, B.: Extraction of user profile based on the hadoop framework. In: 5th International Conference on Wireless Communications, Networking and Mobile Computing, WiCom 2009, pp. 1–6. IEEE (2009)
15. Hyytiä, E., Righter, R., Aalto, S.: Task assignment in a heterogeneous server farm with switching delays and general energy-aware cost structure. Perform. Eval. **75**, 17–35 (2014)

16. Kang, C.W., Abbaspour, S., Pedram, M.: Buffer sizing for minimum energy-delay product by using an approximating polynomial. In: Proceedings of the 13th ACM Great Lakes Symposium on VLSI, pp. 112–115. ACM (2003)

17. Kaxiras, S., Martonosi, M.: Computer architecture techniques for power-efficiency. Synth. Lect. Comput. Archit. **3**(1), 1–207 (2008)

18. Kulkarni, A.P., Khandewal, M.: Survey on hadoop and introduction to YARN. Int. J. Emerg. Technol. Adv. Eng. **4**(5), 82–87 (2014)

19. Rosti, E., Schiavoni, F., Serazzi, G.: Queueing network models with two classes of customers. In: Proceedings Fifth International Symposium on Modeling, Analysis, and Simulation of Computer and Telecommunication Systems, MASCOTS 1997, pp. 229–234. IEEE (1997)

20. Thusoo, A., Sarma, J.S., Jain, N., Shao, Z., Chakka, P., Anthony, S., Liu, H., Wyckoff, P., Murthy, R.: Hive: a warehousing solution over a map-reduce framework. Proc. VLDB Endow. **2**(2), 1626–1629 (2009)

21. Zaharia, M., Borthakur, D., Sarma, J.S., Elmeleegy, K., Shenker, S., Stoica, I.: Job scheduling for multi-user mapreduce clusters. Technical Report UCB/EECS-2009-55, EECS Department, University of California, Berkeley (2009)

A New Modelling Approach to Represent the DCF Mechanism of the CSMA/CA Protocol

Marco Scarpa$^{(\boxtimes)}$ and Salvatore Serrano

Department of Engineering, University of Messina,
Contrada di Dio, 98166 S. Agata, Messina, Italy
{mscarpa,sserrano}@unime.it

Abstract. In this paper, a Markovian agent model is used to represent the behavior of wireless nodes based on CSMA/CA access method. This kind of network was usually modeled by means of bidimensional Markov Chains and more recently using semi-Markov process based models. Both these approaches are based on the assumptions of both full load network and independence of collision probability with respect to retransmission count of each packet. Our model inherently releases the latter hypothesis since it is not necessary to establish a constant collision probability at steady state.

Here, we investigate the correctness of our approach analyzing the throughput of a network based on two IEEE 802.11g nodes when the amount of traffic sent by each one varies. Results have been compared with Omnet++ simulations and show the validity of the proposed model.

1 Introduction

In this paper, we propose a new model for the study of multi-hop Carrier Sense Multiple Access with collision avoidance (CSMA/CA)-based networks. The CSMA/CA protocol is used in a great number of wireless networks to access the medium at MAC layer level. For example the IEEE 802.11 employs a CSMA/CA mechanism with binary exponential backoff (BEB) rules, called Distributed Coordination Function (DCF). DCF defines a basic access method, and an optional four-way handshaking technique, known as request-to-send/clear-to-send (RTS/CTS) method [1]. Even if in this paper we address the basic access mechanism, the approach here presented is promising to be extended in order to take into account the RTS/CTS mechanism. We provide a powerful model based on Markovian agents that accounts for all the exponential backoff protocol details, and allows us to compute throughput performance of DCF for the basic access mechanisms. A more realistic modeling of the mechanism is possible by means of Markovian agents because the relative load condition and retransmission status of each wireless node in the network is taken into account. Practically, using our approach, it is possible to release the key approximation assumed in [2], and subsequent works [3–5] related to the assumption of constant and independent collision probability of a transmitted packet regardless of the number

© Springer International Publishing AG 2017
N. Thomas and M. Forshaw (Eds.): ASMTA 2017, LNCS 10378, pp. 181–195, 2017.
DOI: 10.1007/978-3-319-61428-1_13

of already done retransmissions. In other words, our model will be capable to intrinsically analyze saturated and unsaturated traffic cases [6,7].

The rest of the paper is organized as follows: Sect. 2 presents a summary of the most important papers dealing with modeling of the DCF mechanism; Sect. 3 summarizes the basic access method of DCF; Sect. 4 briefly summarizes the MA theory and introduces the MA model of a wireless node implementing the DCF mechanism; Sect. 5 presents the reference scenario, and the simulation environment we used to validate the MA model; moreover numerical results from both simulation and analytical solution of a MA based scenario are shown. Finally Sect. 6 reports conclusions.

2   Related Work

A great number of different models are introduced for the CSMA-based protocols, most of them, after Bianchi's seminal paper [2], are based on two-dimensional Markov models.

The problem is solved by Bianchi in two distinct steps. In the first step, by means of a Markov model, the behavior of a single wireless node, also called $station$[1], is studied. This model allows to obtain the probability that a station transmits a packet in a generic time slot. In the second step, the throughput is expressed as function of the probability evaluated in the previous step by analyzing the events that can occur within a generic time slot. Two independent processes are considered to model the BEB mechanism: $b(t)$ (the stochastic process representing the backoff time counter for a given station) and $s(t)$ (the stochastic process representing the backoff stage of the station at time t). A restrictive hypothesis is to consider the collision probability p constant regardless of the number of retransmissions that each packet suffers. Assuming the independence of this probability with respect to the number of retransmissions, it is possible to assume a constant value for the same probability p and to model the entire process by means of a bidimensional discrete-time Markov chain.

Recently, a novel Semi-Markov Process (SMP) based model for the single-hop IEEE 802.11 has been proposed to mitigate the complexity of the Bianchi's based model [3]. An SMP is a generalization of Markov chain. It includes a state holding-time that permits to increase the specification of the process. The holding-time of a state i is the amount of time spent before leaving state i. Differently by a traditional Markov process, the future state of a SMP depends not only on the current state but also on its state holding-time. The authors of [3] design the model in three different steps: (1) They build a $(m + 1)$-state Markov Chain. The state i in this chain represents the ith backoff stage of the BEB mechanism. An unsuccessful transmission is modeled by a transition from a lower state i to a higher state $(i + 1)$. A successful transmission is modeled by a transition from any state i to state 0. Loopback transitions are possible only for states 0 and m. (2) The Markov chain defined in step (1) is transformed into an embedded Markov chain. An embedded Markov chain is characterized

---

[1] From this point on, we use the terms $station$ and $wireless\ node$ interchangeably.

by transition probability $P_{ii} = 0$, $\forall i$. The state holding-times of a discrete-time Markov chain are equal to a unit time and are independent of the next state transition. It is possible to consider the sample paths of a SMPs as timed sequences of state transitions. If one watches the process at the times of state transitions, the sample paths of the SMP are identical to those of a Markov chain. Such a process is known as embedded Markov chain. (3) The embedded Markov chain is transformed to an SMP in which the state holding-time for state i will be a random value uniformly selected within the range $(0, 2^i \cdot CWmin)$, for $0 \leq i \leq m$. Once the model is obtained, the authors use the stationary probability distribution of the SMP and the state holding-times to compute the packet transmission probability τ and the saturation throughput in the network. Indeed, the state holding-time for state i in the embedded Markov chain models the backoff interval of backoff stage i and the stationary probability Π_i^s of a SMP represents the fraction of time spent by a wireless node in backoff stage i. The proposed SMP model achieves accurate results with less complexity and computational time with respect to Bianchi's model. An advanced SMP model was proposed in [4]. It calculates the network parameters of single-hop WLANs more accurately. Some other studies also deal with the modeling of CSMA-based protocols in multi-hop scenarios [5,8]. In [9], the authors apply SMP modeling to linear multi-hop networks. The key assumption in all these models is that the probability p that a transmitted packet suffers a collision is the probability that at least one of the $N - 1$ remaining stations transmit in the same time slot. Moreover, if each station transmits a packet with probability τ and it is also assumed the independence of collisions and retransmissions, it is possible to prove that, at steady state, $p = 1 - (1 - \tau)^{N-1}$. In [6], an extension of the model proposed by Bianchi is presented with the aim to evaluate network load in saturation condition taking into account the channel state during the backoff countdown process. Moreover, the authors extend the model to unsaturated traffic cases through an iterative approach which allows to obtain accurate performance metric estimations for a wide range of parameters. Instead in [7], the authors derive analytical expressions for unsaturated network throughput of the IEEE 802.11 DCF using three different schemes: the physical-layer network coding, the traditional nonphysical-layer network-coding, and without network-coding. As mentioned above, in this work we introduce a model that releases all these assumptions and consequently allows to evaluate the behavior of a saturated and an unsaturated network, also considering different traffic loads for each node of the network.

3   The Distributed Coordination Function (Basic Access Method)

In this section, for the sake of completeness, we briefly explain the operation of the basic access method using the DCF mechanism in a IEEE 802.11g WiFi network. In a IEEE802.11g network, when the basic access mechanism is used, a station with packets to transmit monitors the channel activities until it observes the channel is busy. If an idle period equals to a distributed inter-frame space (DIFS) is detected, the station starts the random access mechanism implemented by the backoff period. This latter period is slotted so it is possible to express its length in terms of an integer number of elementary backoff slots. After sensing an idle DIFS, the station initializes a backoff counter. This counter will be decremented by one whenever a time slot expires while the channel is sensed idle. The counter will be stopped when the channel becomes busy due to a transmission of another station and reactivated when the channel will be sensed idle again for more than a DIFS. The station will begin the transmission of the packet when the backoff counter will reach zero. In this manner each station will wait for a random backoff period for an additional deferral time before transmission. If at least two stations decide to start transmission in the same time slot, a collision occurs. When the backoff mechanism starts, the backoff counter is uniformly chosen in the range $[0, CW]$, where CW is the current backoff contention window size. At the first attempt of each packet transmission, CW is set equal to the minimum contention window size CW_{min}. After each unsuccessful transmission (i.e. when a collision occurs), CW is doubled so the next backoff period could be chosen longer. The window will be doubled until the maximum contention window size CW_{max} will be reached. If the window size reaches CW_{max} and other collisions occur, CW shall remain at the value CW_{max}. The backoff contention window size will be reset to CW_{min} value either if a successful attempt to transmit occurs or if the retransmission counter reaches a predefined retry limit (the maximum retransmission number). When the retry limit is reached, the present packet is dropped. If the a packet transmission attempt is successful and the destination station successfully receives the packet, this one responds with an acknowledgment (ACK). The ACK will be followed by a short inter-frame space (SIFS) time during which the channel will be idle. If the transmitting station does not receive the ACK within a specified ACK_Timeout, or it detects the transmission of a different packet on the channel, it decides for collision and reschedules the packet transmission. Furthermore, after the reception of an error packet, a station shall be wait an extended inter-frame space (EIFS) before starting the transmission of a new packet.

In the literature, there have been considerable researches aiming at analyzing the performance of the DCF mechanism.

4   Markovian Agent to Model a DCF Node

In recent years, a new versatile analytical technique [10,11] has emerged whose main idea is to model a distributed system by means of interacting agents. This technique defines each agent through its local properties, but also introduces a mechanism in order to modify its own behavior according to the influence of the interactions with other agents. In this way, the analysis of each agent alone incorporates the effect of the inter-dependencies.

In this section, we briefly recall the basics on Markovian Agents Models and then we describe our proposal for representing the behavior of two interacting wireless stations.

4.1 Markovian Agents

Markovian Agent Models (MAMs) represent systems as a collection of agents scattered over a geographical space, and described by a continuous-time Markov chain (CTMC) where two types of transitions may occur: *local transitions* and *induced transitions*. The former models the internal features of the MA, whereas the latter accounts for interaction with other MAs. When a local transition occurs, an MA can send a message to other MAs. The propagation of messages is regulated by the *perception function* $u(\cdot)$. Depending on the agent position in the space, on the message routing policy, and on the transmittance properties of the medium, this function allows the receiving MA to be aware of the state from which the message was issued, and to use this information to choose an appropriate action. MAs can be scattered over a geographical area $\mathcal{V}$. Agents can be grouped in classes and can share different types of messages.

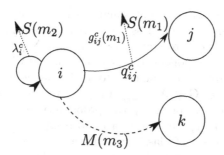

Fig. 1. Graphical representation of MAs

We represent a Markovian agent by exploiting the graphical notation like in Fig. 1. Given an MA of class c, a local transition from state i to state j is drawn as usual with a solid arc and, eventually, the associated rate q_{ij}^c. When a transition happens a message m could be sent with probability $g_{ij}^c(m)$; this event is graphically drawn as a dotted line starting from the transition whose firing sends the message and it is labeled with $S(m)$ to make evident which message is sent. An MA is also able to send a message during the sojourn in a particular state: self-loops are used to this aim; in fact, a self-loop in state i could be used to send a message m at a given rate λ_i^c similarly to what happens during a state transition. It is worth to note that self-loops do not influence local behavior of MAs, like in the usual theory of CTMCs, due to the memoryless property of the exponential distribution; instead they have a role in the evolution of a remote MA receiving the sent message. An example of MA is depicted in Fig. 1, where message m_1 is sent at the occurrence of transition from i to j, and a self-loop is associated with state i emitting message m_2 at rate λ_i^c. Induced transition due to reception of a message is graphically represented with a dashed arc between involved states; in this case the arc is labeled with $M(m)$. As an example, in Fig. 1, a transition from state i to state k, is due to the reception of message m_3.

Formally a *Multiple Agent Class, Multiple Message Type* Markovian Agents Model (MAM) is defined by the tuple:

$$MAM = \{\mathcal{C}, \mathcal{M}, \mathcal{V}, \mathcal{U}, \mathcal{R}\}, \tag{1}$$

where $\mathcal{C} = \{1 \dots C\}$ is the set of agent classes, and $\mathcal{M} = \{1 \dots M\}$ is the set of message types. $\mathcal{V}$ is the finite space over which Markovian Agents are spread, and $\mathcal{U} = \{u_1(\cdot) \dots u_M(\cdot)\}$ is a set of M perception functions (one for each message type). The density of the agents is regulated by functions $\mathcal{R} = \{\xi^1(\cdot) \dots \xi^C(\cdot)\}$, where each component $\xi^c(\mathbf{v})$, with $c \in \mathcal{C}$, counts the number of class c agents deployed in position $\mathbf{v} \in \mathcal{V}$. Since in this work the space is considered discrete, each position could be identified by a *cell* numbered with an integer with respect to some reference system.

Each agent MA^c of class c is defined by the tuple:

$$A^c = \{\mathbf{Q}^c, \mathbf{\Lambda}^c, \mathbf{G}^c(m), \mathbf{A}^c(m), \boldsymbol{\pi}_0^c\}. \tag{2}$$

Here, $\mathbf{Q}^c = [q_{ij}^c]$ is the $n_c \times n_c$ infinitesimal generator matrix of the CTMC that describes the local behavior of a class c agent, and its element q_{ij}^c represents the transition rate from state i to state j (and $q_{ii}^c = -\sum_{j \neq i} q_{ij}^c$). $\mathbf{\Lambda}^c = [\lambda_i^c]$, is a vector of size n_c whose components represent the rates at which the Markov chain reenters the same state: this can be used to send messages with an assigned rate without leaving a state. $\mathbf{G}^c(m) = [g_{ij}^c(m)]$ and $\mathbf{A}^c(m) = [a_{ij}^c(m)]$ are $n_c \times n_c$ matrices that represent respectively the probability that an agent of class c generates a message of type m during a jump from state i to state j, and the probability that an agent of class c accepts a message of type m in state i and immediately jumps to state j. $\boldsymbol{\pi}_0^c$, is a probability vector of size n_c which represents the initial state distribution.

The perception function of a MAM is formally defined as $u_m : \mathcal{V} \times \mathcal{C} \times \mathbb{N} \times \mathcal{V} \times \mathcal{C} \times \mathbb{N} \to \mathbb{R}^+$. The values of $u_m(\mathbf{v}, c, i, \mathbf{v}', c', i')$ represent the probability that an agent of class c, in position $\mathbf{v}$, and in state i, perceives a message m generated by an agent of class c' in position $\mathbf{v}'$ in state i'. Thanks to perception functions, a different instances of agents deployed over the space can interact sending messages one each others. Interactions are technically implemented through the matrix $\mathbf{\Gamma}^c(t, \mathbf{v}, m)$, a diagonal matrix collecting the total rate of received messages m by an agent of class c in position $\mathbf{v}$ (element γ_{ii} stores the value for state i). Matrix $\mathbf{\Gamma}^c(t, \mathbf{v}, m)$ is used to compute the infinitesimal generator matrix of class c agent at position $\mathbf{v}$ at time t: $\mathbf{K}^c(t, \mathbf{v}) = \mathbf{Q}^c + \sum_m \mathbf{\Gamma}^c(t, \mathbf{v}, m)[\mathbf{A}^c(m) - \mathbf{I}]$. The overall Markovian agent model thus evolves according the set of coupled differential equations

$$\frac{d\rho^c(t, \mathbf{v})}{dt} = \rho^c(t, \mathbf{v})\mathbf{K}^c(t, \mathbf{v}) \tag{3}$$

under the initial condition $\rho^c(0, \mathbf{v}) = \xi^c(\mathbf{v})\boldsymbol{\pi}_0^c, \forall \mathbf{v} \in \mathcal{V}, \forall c \in \mathcal{C}$.

As deeply described in [10,11], the main advantage of Markovian agent is that state space complexity is maintained low because dependencies between two agents are modeled through messages instead defining the cross product of their state spaces. Solution method to solve Eq. (3) uses discretization techniques for both time and space and fixed point based algorithms.

4.2   MA Classes in the Model

In the context of DCF, we use two classes of MAs to represent each wireless node: the *Buffer* and the *Backoff* classes. The two classes are depicted in Figs. 2 and 3 respectively.

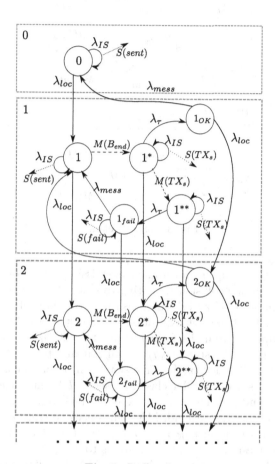

Fig. 2. *Buffer* class

The *Buffer* class models the behavior of wireless node with respect to the transmission of stored frames. It is constituted by $N + 1$ blocks, being N the buffer dimension of the wireless node. A generic block i in the class represents the wireless node buffer containing i frames. When some frames are buffered ($i > 0$), the wireless node tries to transmit one of them at the expiration of the backoff counter. To model this mechanism, we used five states. State i identifies the wireless node with i frames in the buffer waiting for the counter expiration; at the expiration, signaled by the message B_{end} sent by the Backoff class agent, the state changes into i^*. Since each MA of Buffer class sends a TX_s message

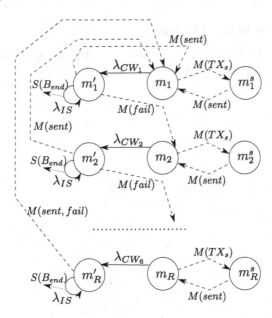

Fig. 3. *Backoff* class

when it is in a state attempting for frame transmission, all the others in the same condition can detect the collision transiting to state i^{**} induced by a perceived TX_s message. The states i^* and i^{**} allow us to discriminate whether the message transmission after the time slot expiration will result in a failure or not. When the transition happens from the former state (to state i_{OK}), the transmission will succeed otherwise (to state i_{fail}) it will fail due to a collision. The rate $\lambda_\tau = \frac{1}{\tau}$ reflects the duration of the time slot τ. In order to correctly signal to other MAs its own state, a Buffer class MA sends TX_s messages through self-loops in states i^* and i^{**}. The state $i-1$ is reached from i_{OK}, decreasing the number of buffered frames; instead, MA comes back to state i when a collision occurs. The sojourn time in i_{OK} and i_{fail} is the time t_{mess} to send a frame, thus we set the corresponding state transitions to $\lambda_{mess} = \frac{1}{t_{mess}}$. Messages $fail$ and $sent$ are used to notify to the Backoff class MA the outcome of transmission attempt. Of course, irrespective of the state, a new packet to be transmitted could arrive from higher layer; to model this event, we connected each state in block i, with $0 < i \leq N-1$, with the equivalent in block $i+1$. We denote the frame arrival rate with λ_{loc}.

Block 0 has a different structure since it represents the empty buffer thus no transmission attempts are done and one state only (the state 0) completely represent the system. The message sent in 0 is used to trigger the start of Backoff class MA in the case of countdown expiration.

Based on this description, matrices storing transition rates and self-loop rates can be easily built. The *Buffer* class initial probability vector set to 1.0 the probability to be in state 0. Generating matrices and acceptance matrices are

set in such a way the described messages are sent and received with probability
1.0. We do not write their complete structure for lack of space.

The Backoff class (Fig. 3) has the task of regulating the starting of transmissions of *Buffer* class agents by sending the B_{end} message. This is implemented with three states denoted in the following as m_i, m_i' and m_i^s. State m_i represents the system waiting for the backoff counter expiration in the contention window CW_i; when the contention timeout expires the MA transits to m_i'. The transition rate from m_i to m_i' is computed taking into consideration the contention windows at i-th attempt as follows:

$$\lambda_{CW_i} = \begin{cases} \frac{1}{2^{i-1}CW_{min}} & 2^{i-1}CW_{min} < CW_{max} \\ \frac{1}{CW_{max}} & otherwise \end{cases} \tag{4}$$

A B_{end} message is emitted in the states m_i' signaling to the perceiving MAs that the contention timeout expired.

All the other transitions in this class are induced by other MAs. In particular, when a *Backoff* class MA is in m_i' it can transit in either m_{i+1} or m_1 if the corresponding *Buffer* MA failed in transmitting the frame or not, respectively. These events are considered by perceiving the messages *fail* and *sent* emitted by the MA modeling the frame transmission. In presence of more *Buffer* class MAs, only the messages generated by the MA in the same location of the Backoff MA are perceived. Instead transitions back and forth to m_i^s are due to messages emitted by Backoff class MAs in different locations of the receiving Buffer class MA. In fact a TX_s message has to be perceived if emitted by some transmitting stations, different by the actually considered, because it acquired the channel; thus the countdown is stopped (transition to m_i^s) and it will be resumed (transition back to m_i) at the reception of *sent* message.

A slightly different behavior is implemented in the state m_R', where a reception of *fail* message reset the backoff algorithm to m_1 because the maximum number of retries has been reached.

As in the previous case, based on this description, matrices storing transition rates and self-loop rates can be easily built. The *Backoff* class initial probability vector set to 1.0 the probability to be in state m_1. Generating matrices and acceptance matrices are set in such a way the described messages are sent and received with probability 1.0. We do not write their complete structure for lack of space.

4.3   The Implemented MAM

In order to define the complete model we have to define all the quantities in (1).

In the case considered in this paper, the space $\mathcal{V}$ is a very little rectangular area composed of two cells. This choice is due to the fact that, in this work, we focus on correctly modeling the interaction mechanism and thus only two transmitting stations are considered. Due to this reason, the space $\mathcal{V}$ is simply defined by two positions and we have $\mathcal{V} = \{0, 1\}$.

The set $\mathcal{C}$ is defined by the two class introduced in Sect. 4.2: for the sake of simplicity, we will denote with "u" and "a" the *Buffer* and *Backoff* class respectively. Thus $\mathcal{C} = \{u, a\}$, whereas $\mathcal{M} = \{TX_s, B_{end}, fail, sent\}$.

Since each wireless node is modeled by a couple of MAs (one *Buffer* MA and one *Backoff* MA) and we considered two wireless nodes in the space, four interacting MAs are in the model. From these considerations, the set $\mathcal{R}$ easily derives being $\xi^u(0) = \xi^a(0) = \xi^u(1) = \xi^a(1) = 1$.

Perception functions $u_m(\cdot)$, with $m \in \mathcal{M}$, are specified in Table 1 where only the non null values are written. In this paper, we assumed that when a message is sent to a given target destination it is perceived, thus all the values of $u_m(\cdot)$ in the table are equal to 1.0. Perception functions reflect the interaction among MA classes as described in Sect. 4.2.

Table 1. MAM perception functions

Message	v	Class	State	v'	Class	State
		Perceiving MA			Receiving MA	
B_{end}	0	u	$i, 1 \leq i \leq N$	0	a	$m'_j, 1 \leq j \leq R$
B_{end}	1	u	$i, 1 \leq i \leq N$	1	a	$m'_j, 1 \leq j \leq R$
TX_s	0	u	$i^*, 1 \leq i \leq N$	1	u	$i^*, 1 \leq i \leq N$
TX_s	0	u	$i^*, 1 \leq i \leq N$	1	u	$i^{**}, 1 \leq i \leq N$
TX_s	0	a	$m_j, 1 \leq j \leq R$	1	u	$i^*, 1 \leq i \leq N$
TX_s	0	a	$m_j, 1 \leq j \leq R$	1	u	$i^{**}, 1 \leq i \leq N$
TX_s	1	u	$i^*, 1 \leq i \leq N$	0	u	$i^*, 1 \leq i \leq N$
TX_s	1	u	$i^*, 1 \leq i \leq N$	0	u	$i^{**}, 1 \leq i \leq N$
TX_s	1	a	$m_j, 1 \leq j \leq R$	0	u	$i^*, 1 \leq i \leq N$
TX_s	1	a	$m_j, 1 \leq j \leq R$	0	u	$i^{**}, 1 \leq i \leq N$
$fail$	0	a	$m'_j, 1 \leq j \leq R$	0	u	$i_{fail}, 1 \leq i \leq N$
$fail$	1	a	$m'_j, 1 \leq j \leq R$	1	u	$i_{fail}, 1 \leq i \leq N$
$sent$	0	a	$m'_j, 1 \leq j \leq R$	0	u	$i, 0 \leq i \leq N$
$sent$	0	a	$m^s_j, 1 \leq j \leq R$	1	u	$i, 0 \leq i \leq N$
$sent$	1	a	$m'_j, 1 \leq j \leq R$	1	u	$i, 0 \leq i \leq N$
$sent$	1	a	$m^s_j, 1 \leq j \leq R$	0	u	$i, 0 \leq i \leq N$

We note that the use of MAs makes simple the extension to a more complex scenario with N nodes: it is enough to deploy a *Buffer* MA and a *Backoff* MA for each wireless node and appropriately define the perception functions for each pair of deployed MAs. Since the purpose of this paper is to show the usefulness and the advantages of Mobile Agent modeling paradigm in representing interacting wireless stations, we consider only a two wireless nodes scenario leaving the study of more complex scenarios as future research.

5   Model Validation and Results

In this preliminary work, we referred to a simple IEEE 802.11g wireless network to evaluate the accuracy of our DSF model; we take into account the network throughput by varying the network load from a lightweight to a saturated one. The IEEE 802.11g operates in the 2.4 GHz ISM band and provides a maximum raw data throughput of 54 Mbps, although this translates to a real maximum throughput of 24 Mbps. In this simple scenario, we set two wireless nodes transmitting an UDP flow one each other. We did not consider any kind of noise or signal interference neither signal attenuation. The flows are characterized by an exponentially distributed UDP payload size. We evaluated model results using both an average value of 100 bytes and an average value of 1000 bytes for the UDP payload size. Moreover, we considered an exponentially distributed generation of information units at the application level.

According to the bidirectional traffic flow in each node, it is possible to evaluate the mean load of the network R, expressed in bps, by using the following relation

$$R = \frac{2 \cdot L}{g} \tag{5}$$

where L is the average frame length expressed in bits ($L = C \cdot t_{mess}$, with $C = 54$ Mbps the maximum raw data rate) and g is the average value of the intrapacket gap (IPG) ($g = \frac{1}{\lambda_{loc}}$). To validate our model, we also simulated the two interacting wireless nodes. We used "Omnet++" as a network simulator. For simulation purpose, we put a 802.11 g access point and a 802.11 g wireless host in a predefined squared area and we imposed a "stationary mobility type". Each device has been equipped with a NIC having a MAC and PHY using the parameters showed in Table 2. Accordingly, the simulator never uses the Request To Send/Clear To Send (RTS/CTS) mechanism because we imposed the maximum length of each packet to 1500 bytes, i.e. the Ethernet Maximum Transfer Unit (MTU). Moreover, considering the PHY parameters and the distance of the nodes, each transmitted packet will be received without errors due to the strictly high SNR. In this conditions only collisions can disrupt the reception of a packet. Finally, the UDP IPG (g) has been assumed to be an exponentially distributed random variable with mean value set accordingly to the desired load of the network. We performed 50 simulation runs for each evaluated point

Table 2. MAC & PHY parameters

Control bit rate	54 Mbps	Basic bit rate	54 Mbps
Max queue size	14 packets	RTS threshold	3000 bytes
Contention window (min size)	31 time slots	Retry limit	7 times
Contention window (max size)	255 time slots	Tx Power	20 mW
Receiver sensitivity	–85 dBm	Noise power	–192 dBm
Distance	10 m		

(a specified network load) each lasting 12 s. Confidence intervals obtained is very narrow, thus we did not show them in the related graphs.

Derivation of network throughput from the MAM model is obtained considering the probability a transmission successfully complete. This is easily evaluated considering the states i_{OK} of the *Buffer* class MAs as follows:

$$Th_f = \sum_{i=1}^{N} (P[M^u(0) = i_{OK}] + P[M^u(1) = i_{OK}]) \lambda_{mess} \tag{6}$$

where $M^u(0)$ $(M^u(1))$ denotes the Buffer class MA in position 0 (1). Th_f gives the network throughput in terms of $frame/s$, from which we computed the equivalent UDP Throughput expressed in Mbps according to the frame size considered in the experiments.

Figure 4 shows the throughput obtained using simulation and agent model as a function of the UDP network load when the packet size has an average size of 100 bytes. For network load greater then 2 Mbps the model gives a constant throughput of 2.21 Mbps while the results of the simulations are slightly higher than this value for load in the range between 2 Mbps and 16 Mbps, they become essentially identical in the range between 16 Mpbs and 20 Mbps and slightly lower for higher load. We have almost the same results for a UDP load of 1.6 Mbps obtaining a throughput of 1.595 Mbps using the simulator and of 1.584 Mbps using the MA model (the percentage absolute error is equal to 0.68%). The maximum value of percentage absolute error (17.19%) was obtained with a load of 32 Mbps when the analytical model gives a throughput higher than about 0.32 Mbps with respect to that obtained by simulator.

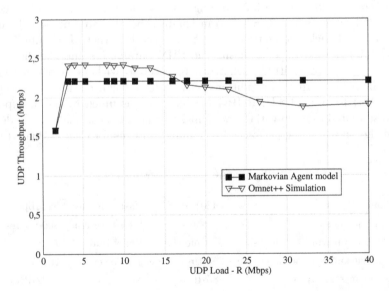

Fig. 4. UDP Throughput vs UDP Load using packets of average size equal to 100 bytes.

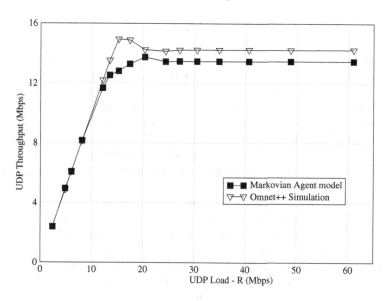

Fig. 5. UDP Throughput vs UDP Load using packets of average size equal to 1000 bytes.

Figure 5 shows the throughput obtained using simulation and MA model as a function of the UDP network load when packets have an average size of 1000 bytes. In this latter case, we obtained more accurate results. For load slower than 10 Mbps, the throughput derived from the MA model is essentially equal to that obtained with simulation (the maximum value of the percentage absolute error was 1.66%). When the load grows up over 20 Mbps the throughput saturate at 14.45 Mbps for the MA model and slightly oscillate around 14.22 Mbps for the simulations. In this latter condition the maximum value of the percentage absolute error is 5.53%. In the range between 10 Mbps and 20 Mbps the throughput obtained by the MA model presents the maximum deviation with respect to the results obtained by the simulation growing up to a percentage absolute error approximately equal to 14%.

The percentage absolute error has been computed as follows:

$$|E_\%| = \frac{|\mathrm{Th}_{\mathrm{model}} - \mathrm{Th}_{\mathrm{simulator}}|}{\mathrm{Th}_{\mathrm{simulator}}} \cdot 100 \qquad (7)$$

and we obtained corresponding value for comparison using linear interpolation (the actual load of the simulations is always different from the theoretical load used to obtain the throughput using the model).

It is worth to note that in the Figs. 4 and 5 we do not show the confidence intervals because they were very small: the larger confidence interval we obtained is 0.08 Mbps using a confidence of 99%; thus we do not show confidence intervals in the graphs because they are very narrow and difficult to read.

6   Conclusions and Future Work

In this paper, we introduced a new performance model of CSMA/CA based networks. Specifically Markovian agents were used to model the behavior of the protocol. In this preliminary work, we focused on a simple network scenario with 2 nodes. In the model, we neglected the presence of interference and noise and, accordingly, we set a very high SNR at the receiver in our simulations. Results obtained in terms of throughput of UDP traffic varying the network load confirmed the goodness of the approach. The results were compared with that obtained by Omnet++ simulations and using both small and large packets, experiencing very small differences.

One of the most important strengths of this proposal is the possibility to release the hypothesis of "loaded network" used in the Bianchi's based model where the collision probability is assumed constant and independent of the number of suffered retransmission. Using the proposed approach, it is instead possible to load each wireless node of the network with a different quantity of traffic and, anyway, it inherently permits to link the collision probability with the retransmission count for each wireless node without imposing a constant value.

In the future, we intend to extend this work considering a more complex scenario taking into account a greater number of wireless nodes (in order to verify model scalability), the presence of interference and/or noise (it will be possible by tuning the perception function of the messages accordingly to the distance and/or signal to noise ratio), and a multi hop (chained) scenario in which each wireless node generates its own traffic but also route traffic coming from its neighbors.

Finally, it is noteworthy to consider that the introduced Markovian agent can also be used to model wireless network using the DCF approach but different of the IEEE 802.11g. A possible candidate we intend to analyze in the future is a wireless sensor network based on the IEEE 802.15.4 protocol.

References

1. IEEE Std 802.11-2012: IEEE Standard for Information technology-telecommunications and information exchange between systems Local and metropolitan area networks-Specific requirements part 11: wireless LAN Medium Access Control (MAC) and Physical Layer (PHY) Specifications, (Revision of IEEE Std 802.11-2007), pp. 1–2793, March 2012
2. Bianchi, G.: Performance analysis of the IEEE 802.11 distributed coordination function. IEEE J. Sel. Areas Commun. **18**(3), 535–547 (2000)
3. Kadiyala, M.K., Shikha, D., Pendse, R., Jaggi, N.: Semi-Markov process based model for performance analysis of wireless LANs. In: 2011 IEEE International Conference on Pervasive Computing and Communications Workshops (PERCOM Workshops), pp. 613–618, March 2011
4. Wang, H., Kang, G., Huang, K.: An advanced Semi-Markov process model for performance analysis of wireless LANs. In: 2012 IEEE Vehicular Technology Conference (VTC Fall), pp. 1–5, September 2012

5. Prasad, Y.R.V., Pachamuthu, R.: Analytical model of adaptive CSMA-CA MAC for reliable and timely clustered wireless multi-hop communication. In: 2014 IEEE World Forum on Internet of Things (WF-IoT), pp. 212–217, March 2014
6. Felemban, E., Ekici, E.: Single hop IEEE 802.11 DCF analysis revisited: accurate modeling of channel access delay and throughput for saturated and unsaturated traffic cases. IEEE Trans. Wirel. Commun. **10**(10), 3256–3266 (2011)
7. Lin, S., Fu, L.: Unsaturated throughput analysis of physical-layer network coding based on IEEE 802.11 distributed coordination function. IEEE Trans. Wirel. Commun. **12**(11), 5544–5556 (2013)
8. Tadayon, N., Wang, H., Chen, H.H.: Performance analysis of distributed access multihop poisson networks. IEEE Trans. Veh. Technol. **63**(2), 849–858 (2014)
9. Stajkic, A., Buratti, C., Verdone, R.: Modeling multi-hop CSMA-based networks through semi-Markov chains. In: 2015 International Wireless Communications and Mobile Computing Conference (IWCMC), pp. 520–525, August 2015
10. Gribaudo, M., Cerotti, D., Bobbio, A.: Analysis of on-off policies in sensor networks using interacting markovian agents. In: Sixth Annual IEEE International Conference on Pervasive Computing and Communications, 2008. PerCom 2008, pp. 300–305, March 2008
11. Bruneo, D., Scarpa, M., Bobbio, A., Cerotti, D., Gribaudo, M.: Markovian agent modeling swarm intelligence algorithms in wireless sensor networks. Performance Evaluation **69**(3–4), 135–149 (2012). Selected papers from ValueTools 2009. http://www.sciencedirect.com/science/article/pii/S0166531611000137

The Mean Drift: Tailoring the Mean Field Theory of Markov Processes for Real-World Applications

Mahmoud Talebi[1(✉)], Jan Friso Groote[1], and Jean-Paul M.G. Linnartz[2]

[1] Department of Mathematics and Computer Science,
Eindhoven University of Technology, Eindhoven, The Netherlands
{M.Talebi,J.F.Groote}@tue.nl
[2] Department of Electrical Engineering, Eindhoven University of Technology,
Eindhoven, The Netherlands
J.P.Linnartz@tue.nl

Abstract. The statement of the mean field approximation theorem in the mean field theory of Markov processes particularly targets the behaviour of population processes with an unbounded number of agents. However, in most real-world engineering applications one faces the problem of analysing middle-sized systems in which the number of agents is bounded. In this paper we build on previous work in this area and introduce the mean drift. We present the concept of population processes and the conditions under which the approximation theorems apply, and then show how the mean drift can be linked to observations which follow from the propagation of chaos. We then use the mean drift to construct a new set of ordinary differential equations which address the analysis of population processes with an arbitrary size.

Keywords: Markov chains · Population processes · Mean field approximation · Propagation of chaos

1 Introduction

Population processes are stochastic models of systems which consist of a number of similar agents (or particles) [1]. When the impact of each agent on the behaviour of the system is similar to other agents, it is said that the population process is a mean field interaction model (or is symmetric) [2]. Mean field approximation refers to the continuous, deterministic approximations of the stochastic behaviour of such processes, when the number of agents grows very large. These approximations were first proposed for several concrete cases in various areas of study e.g., from as early as the 18th century in population biology, where models such as the SIR equations are used to describe the dynamics of epidemics [3].

The authors would like to thank Erik de Vink, Mieke Massink, Tjalling Tjalkens and Ulyana Tikhonova for their constructive comments and helpful remarks.

N. Thomas and M. Forshaw (Eds.): ASMTA 2017, LNCS 10378, pp. 196–211, 2017.
DOI: 10.1007/978-3-319-61428-1_14

Since then, general theorems have been proven which show the convergence of the behaviour of population processes to solutions of differential equations. The proofs follow roughly the same steps which generally rely on Grönwall's lemma and martingale inequalities [4]. One of the first generalized approximation theorems was given by Kurtz [5]. The theory gives conditions which define a family of such models called density-dependent population processes, and finds their deterministic approximations by a theorem which is generally called the law of large numbers for standard Poisson processes [6].

The mean field theory of Markov processes is increasingly being applied in the fields of computer science and communication engineering. In the field of communication engineering and starting with Bianchi's analysis of the IEEE 802.11 DCF protocol [7], much research has focused on discussing the validity of the so-called decoupling assumption in this analysis. Several general frameworks have also been proposed which target the analysis of computer and communication systems [2,8]. In the field of computer science the initial application of the approximations was intuitively motivated by methods such as fluid and diffusion approximations of queueing networks [9]. These have resulted in the development of methods and tools to automate the analysis of mean field models, with extensive progress in the context of the stochastic process algebra PEPA [10,11]. However, still a large family of models are deemed unsuitable for this fluid approximation analysis, since they often lead to demonstrably inaccurate approximations [10,12]. This calls for revisiting the fundamental roots of the theory. Such an approach has been taken in [13] where the authors use a set of extended diffusion approximations to derive precise approximations of stochastic Petri Net models.

We identify the current challenge as the problem of analysing middle-sized systems: systems which are so large that they suffer from *state space explosion*, but not large enough such that they can be accurately analysed by common approximation methods. In this paper we focus on the evaluation of these middle-sized systems, and the most important contribution is the introduction of Poisson mean of intensities (Eq. (11)), and the way they relate to the approximation theorem based on the idea of propagation of chaos.

Through Eq. (11) we express the idea that occupancy measures can be seen as Poisson arrival rates. Based on this observation we propose employing the concept of the *Poisson mean* of the drift to build the set of ordinary differential equations in (12) when dealing with bounded systems. To provide further proof for the consistency of our observation with respect to already established results in the mean field theory, we show that the drift and the mean drift are equivalent in the limit (Theorem 2).

Besides explaining the idea of the mean drift, we also take a slightly different approach in performing the frequent time and probability scalings in building population processes, by systematically removing the dependency of scaling factors on the size of the system.

The rest of this text is organized as follows. Section 2 describes the family of mean field interaction models. In Sect. 3 we introduce the drift of a population process and the derivation of their deterministic approximations, which provide the basis for stating the main results of this work. Finally, in Sect. 4 we present the idea of propagation of chaos and define the concept of mean drift.

2   Population Processes and the Mean Field Model

In this section, we present the stochastic model of a system and introduce mean field interactions models. For the most part our notation agrees with [2]. The list of objects appearing in this paper are given in Table 1.

Table 1. Table of objects and their short description.

$T = \mathbb{N}$	Points corresponding to local time-slots
$T_G \subseteq \mathbb{Q}_{\geq 0}$	Points on the real line corresponding to global time-slots
$D \in \mathbb{N}_{\geq 1}$	Time resolution = number of global time-slots in a unit interval
$\epsilon = \frac{1}{D}$	Length of a global time-slot
$N \in \mathbb{N}_{\geq 1}$	System size = number of agents
$\mathcal{S} = \{1, \ldots, I\}$	State space of agents, with $I \in \mathbb{N}$ states
$\left\{ X_i^{(N)}(t) : t \in T \right\}$	Process corresponding to agent i, with $i \in \{1, \ldots, N\}$
K_i	Transition map of $X_i^{(N)}(t)$
$\left\{ \hat{X}_i^{(N)}(t) : t \in T_G \right\}$	Modified process corresponding to agent i
$\hat{K}_i$	Transition map of $\hat{X}_i^{(N)}(t)$
$\{ Y^{(N)}(t) : t \in T_G \}$	Process for the system of N agents, on $\mathcal{S}^N$
$\mathcal{K}^{(N)}$	Transition map of $Y^{(N)}(t)$
Δ	Set of occupancy measures
$\{ M^{(N)}(t) : t \in T_G \}$	Normalised population process on $\Delta^{(N)} \subset \Delta$
$P_1^{(N)}$	Transition map of the agent model $\left\{ \left(\hat{X}_1^{(N)}, M^{(N)}(t) \right) : t \in T_G \right\}$
$P_{s,s'}^{(N)}$	Agent transition map, with $s, s' \in \mathcal{S}$
$Q_{s,s'}^{(N)}$	Infinitesimal agent transition map, with $s, s' \in \mathcal{S}$
$\{ \bar{M}^{(N)}(t) : t \in \mathbb{R}_{\geq 0} \}$	Normalised population process with continuous paths
$\{ W^{(N)}(t) : t \in T_G \}$	Object (agent) state-change frequency in interval $(t - \epsilon, t]$
$\hat{F}^{(N)}$	Expected instantaneous change in system state
$F^{(N)}$	Drift of the normalized population process
$\Phi \subseteq \{ g : \mathbb{R}_{\geq 0} \to \Delta \}$	Set of deterministic approximations
F^*	The limit of the sequence of drifts $\{ F^{(N)} \}$
ρ_N	The probability measure induced by $Y^{(N)}(t)$
$\bar{F}_{s,s'}^{(N)}$	The Poisson mean of intensity from s to s' with $s, s' \in \mathcal{S}$
$\bar{F}^{(N)}$	The mean drift of the normalised population process

2.1   Agent Processes and the Clock Independence Assumption

Let the set $T = \mathbb{N}$ be discrete and let parameter $N \in \mathbb{N}_{\geq 1}$ be the *system size*. The elements of T are called time-slots. Let $\mathcal{S} = \{1, \ldots, I\}$ be a finite set of states. For $i \in \{1, \ldots, N\}$, let $\left\{ X_i^{(N)}(t) : t \in T \right\}$ be $\mathcal{S}$-valued discrete-time time-homogeneous Markov chains (DTMCs). Each stochastic process $X_i^{(N)}(t)$ describes the behaviour of agent i in the system with N agents.

Take each process $X_i^{(N)}(t)$ to be described by a transition map $K_i : \mathcal{S}^N \times \mathcal{S} \to [0,1]$. In each time-slot (indexed by members of T), the process chooses the next state $s \in \mathcal{S}$ with probability $K_i(\boldsymbol{v}, s)$, where the vector of states $\boldsymbol{v} \in \mathcal{S}^N$ is the state of the entire system (including agent i's current state).

We assume that processes have independent time-slots, which occur at the same rate over sufficiently long intervals of time. This *clock independence* assumption allows us to embed the discrete-time description of agents' behaviours in a continuous-time setting. For a discussion on the approximation of systems with simultaneous update (or synchronous DTMCs [11]) refer to [8,11].

Formally, the independence of time-slots can be stated as follows. For two processes i and i' where $i \neq i'$, if process i does a transition in an instant of time then process i' almost never does a transition simultaneously. Technically, the clock independence assumption is realized as follows. Let $D \in \mathbb{N}^+$ be the time resolution, and let $\epsilon \in \mathbb{Q}_{\geq 0}$ be a positive rational number (a probability) defined as $\epsilon = \frac{1}{D}$. Let $T_G \subseteq \mathbb{Q}_{\geq 0}$ be the countable set: $T_G = \{0, \epsilon, 2\epsilon, \ldots\}$. We call the set T_G the *system* or *global time*, as opposed to the *agent* or *local time* T.

Let the probability of an agent doing a transition in a time-slot be ϵ. In this new setting, for $1 \leq i \leq N$ define stochastic processes $\left\{ \hat{X}_i^{(N)}(t) : t \in T_G \right\}$, each with transition maps $\hat{K}_i : \mathcal{S}^N \times \mathcal{S} \to [0,1]$, such that for all $\boldsymbol{v} \in \mathcal{S}^N$ and $s \in \mathcal{S}$:

$$\hat{K}_i(\boldsymbol{v}, s) = \begin{cases} \epsilon \, K_i(\boldsymbol{v}, s) & \text{if } s \neq \boldsymbol{v}_i, \\ (1 - \epsilon) + \epsilon \, K_i(\boldsymbol{v}, s) & \text{if } s = \boldsymbol{v}_i. \end{cases}$$

In the new setting, let E be the event that agent i does a transition in a time-slot, and E' be the event that agent $i' \neq i$ does a transition exactly in the same time-slot in T_G. Then by independence of agent transition maps: $\mathbb{P}\{E'|E\} = \mathbb{P}\{E'\} = \epsilon$. Observe that the clock independence assumption is satisfied as $D \to \infty$ (i.e., $\epsilon \to 0$). A similar approach can be taken to relate the behaviour of agents specified using CTMCs as well, which results in an approximation theorem which closely resembles ours (see [14]).

Let $Y^{(N)}(t) = \left(\hat{X}_1^{(N)}(t), \ldots, \hat{X}_N^{(N)}(t) \right)$ be a stochastic process. The process $Y^{(N)}(t)$ represents the behaviour of the entire system, and is also a time-homogeneous discrete-time Markov process with transition map $\mathcal{K}^{(N)} : \mathcal{S}^N \times \mathcal{S}^N \to [0,1]$ in which for $\boldsymbol{v}, \boldsymbol{v}' \in \mathcal{S}^N$:

$$\mathcal{K}^{(N)}(\boldsymbol{v}, \boldsymbol{v}') = \prod_{i=1}^{N} \hat{K}_i(\boldsymbol{v}, \boldsymbol{v}'_i).$$

2.2   Mean Field Interaction Models and Population Processes

We now introduce mean field interaction models [2], which comprise the class of processes $Y^{(N)}(t)$ for which we find the mean field approximations.

Let $\pi : \{1, \dots, N\} \to \{1, \dots, N\}$ be a permutation over the set $\{1, \dots, N\}$. For a vector $\boldsymbol{v} = (s_1, \dots, s_N)$ define $\pi(\boldsymbol{v})$ as:

$$\pi(\boldsymbol{v}) = \left(s_{\pi(1)}, \dots, s_{\pi(N)}\right).$$

Definition 1 (*Mean Field Interaction Models* [2]). *Let $Y^{(N)}(t)$ be the process defined earlier, and let π be any permutation over the set $\{1, \dots, N\}$. If for all $\boldsymbol{v}, \boldsymbol{v}' \in \mathcal{S}^N$,*

$$\mathcal{K}^{(N)}(\boldsymbol{v}, \boldsymbol{v}') = \mathcal{K}^{(N)}(\pi(\boldsymbol{v}), \pi(\boldsymbol{v}'))$$

holds, $Y^{(N)}(t)$ is called a mean field interaction model with N objects.

It follows from the above definition that entries in $\mathcal{K}^{(N)}$ may depend on the number of agents in each state, but not on the state of a certain agent. Let $\Delta = \{\boldsymbol{m} \in \mathbb{R}^I : \sum_{s \in \mathcal{S}} \boldsymbol{m}_s = 1 \wedge \forall s. \boldsymbol{m}_s \geq 0\}$ be a set of vectors, which we call the set of *occupancy measures*. For a system of size N, take the countable subset $\Delta^{(N)} = \{\boldsymbol{m} \in \mathbb{R}^I : \sum_{s \in \mathcal{S}} \boldsymbol{m}_s = 1 \wedge \forall s. N\boldsymbol{m}_s \in \mathbb{N}\}$. The set $\Delta^{(N)} \subset \Delta$ is an alternative representation of the state space of the system, in which for $\boldsymbol{m} \in \Delta^{(N)}$ and $i \in \mathcal{S}$ the value $\boldsymbol{m}_i$ expresses the proportion of agents that are in state i. For a mean field interaction model we define the *normalized population process* $\{M^{(N)}(t) : t \in T_G\}$ on $\Delta^{(N)}$ such that for $s \in \mathcal{S}$:

$$M_s^{(N)}(t) = \frac{1}{N} \sum_{1 \leq n \leq N} \mathbb{1}\left(\hat{X}_n^{(N)}(t) = s\right), \tag{1}$$

where $\mathbb{1}$ is an indicator function. Using the fact that $Y^{(N)}(t)$ is a mean field interaction model, it is possible to move back and forth between processes $Y^{(N)}(t)$ and $M^{(N)}(t)$.

We define the behaviour of an agent in the context of a population by the process $\left\{\left(\hat{X}_1^{(N)}(t), M^{(N)}(t)\right) : t \in T_G\right\}$, called the *agent model* with the transition map $P_1^{(N)} : \mathcal{S} \times \Delta \times \mathcal{S} \times \Delta \to [0, 1]$. Then for $s, s' \in \mathcal{S}$, the marginal transition probability that an agent moves from state s to state s' in the context $\boldsymbol{m} \in \Delta^{(N)}$ is defined as:

$$P_{s,s'}^{(N)}(\boldsymbol{m}) = \sum_{\boldsymbol{m}' \in \Delta} P_1^{(N)}(s, \boldsymbol{m}, s', \boldsymbol{m}').$$

For each $s, s' \in \mathcal{S}$ where $s \neq s'$, the expected proportion of the agents that are in state s at time t and move to state s' over a unit time interval $[t, t+1)$ is $D P_{s,s'}^{(N)}(\boldsymbol{m})$. By taking the clock independence assumption into account, for $s \neq s'$ define the functions $Q_{s,s'}^{(N)} : \Delta^{(N)} \to \mathbb{R}_{\geq 0}$ which for $\boldsymbol{m} \in \Delta^{(N)}$ satisfy:

$$Q_{s,s'}^{(N)}(\boldsymbol{m}) = \lim_{D \to \infty} D P_{s,s'}^{(N)}(\boldsymbol{m}).$$

Note that due to the construction of the probabilities $P_{s,s'}^{(N)}(\boldsymbol{m})$, for $s \neq s'$ the limit always exists. The mapping $Q^{(N)}$ is called the infinitesimal agent transition map, which can be interpreted as a transition rate matrix, meaning that for an agent in state s the time until it moves to state s' converges to an exponentially distributed random variable with mean: $\frac{1}{Q_{s,s'}^{(N)}}$.

Despite the fact that time instants in the set T_G are discrete, we deem it necessary to observe the population process $\{M^{(N)}(t) : t \in T_G\}$ at continuous times $t \in \mathbb{R}_{\geq 0}$. Based on $M^{(N)}(t)$ we define a new stochastic process $\{\bar{M}^{(N)}(t) : t \in \mathbb{R}_{\geq 0}\}$ with new sample paths which are right-continuous functions with left limits (càdlàgs). For $t \in \mathbb{R}_{\geq 0}$ the process $\bar{M}^{(N)}(t)$ satisfies:

$$\bar{M}^{(N)}(t) = M^{(N)}\left(\epsilon\lfloor Dt\rfloor\right).$$

In this and the following sections we use a running example inspired by radio communication networks, to illustrate how one derives the mean field approximation of a system. In the example of this section we start from an informal specification of the network behaviour, and derive the transition map of the corresponding normalized population process.

Example 1. Consider a network of N nodes (agents) operating on a single shared channel. The network is saturated, meaning that all the nodes always have messages to transmit. A node can be in one of the states $S = \{1, 2\}$. In state 1 a node is waiting, and with probability p_1 decides to transmit a message. All communications start with the transmission of a message, and a successful communication is then marked by the receipt of an acknowledgement, whereas a failed communication ends in a timeout. Both cases occur in the space of a single time-slot. If the communication succeeds, the node will remain in state 1 and wait to transmit the next message and if it fails, the node moves to state 2 in which it tries retransmitting the message. A node in state 2 retransmits the message with probability p_2. The node then essentially behaves in the same way as in state 1.

The probability of success depends on the number of nodes currently using the channel, as follows. If $n \in \mathbb{N}$ nodes are using the channel, then the success probability of each participating communication is:

$$p_s(n) = 2^{-n}, \tag{2}$$

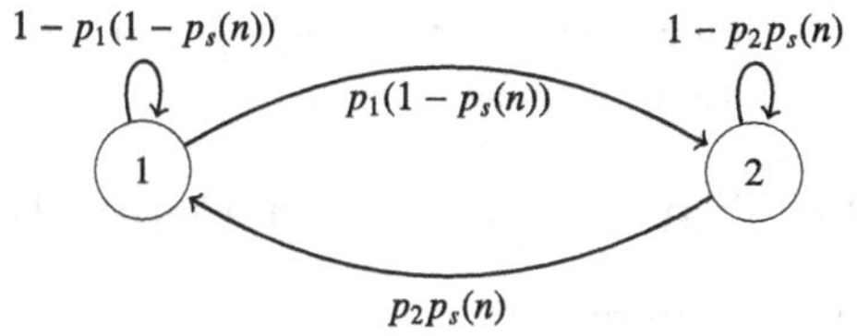

Fig. 1. The behaviour of a node in Example 1. The number of transmitting nodes is n.

that is, the channel degrades in quality exponentially as the number of active nodes increases. A diagram representing the behaviour of each node is given in Fig. 1.

Let $v \in S^N$ be the state of the network, with n_1 nodes in state 1 and $n_2 = N - n_1$ nodes in state 2. Let $tr_1(v)$ be the total number of nodes that are in state 1 which decide to transmit a message in system state v, then $tr_1(v)$ is a binomial random variable with distribution $B(n_1, p_1)$. In a similar fashion $tr_2(v)$, the total number of nodes in state 2 which decide to transmit a message in system state v is $B(n_2, p_2)$ distributed. Then a communication in state v will succeed with probability $p_s(tr_1(v) + tr_2(v))$.

Let $s \in S$ be the next state of a node i $(1 \leq i \leq N)$, based on the description above the transition matrix for this node is:

$$K_i(v, s) = \begin{pmatrix} 1 - p_1 + p_1 p_s(tr_1(v) + tr_2(v)) & p_1 - p_1 p_s(tr_1(v) + tr_2(v)) \\ p_2 p_s(tr_1(v) + tr_2(v)) & 1 - p_2 p_s(tr_1(v) + tr_2(v)) \end{pmatrix}$$

where the row is determined by the element v_i (current state) and the column by s.

We use the clock independence assumption to compose the population process. To extend this assumption to the description of our radio network, we implicitly assume that the duration of message transmission is exponentially distributed, i.e., since the transitions are memoryless, the sojourn time of individuals in states is exponentially distributed. The modified transition matrix for node i is:

$$\hat{K}_i(v, s) = \begin{pmatrix} 1 - p_1 \epsilon (1 - p_s(tr_1(v) + tr_2(v))) & p_1 \epsilon (1 - p_s(tr_1(v) + tr_2(v))) \\ p_2 \epsilon p_s(tr_1(v) + tr_2(v)) & 1 - p_2 \epsilon p_s(tr_1(v) + tr_2(v)) \end{pmatrix}$$

In which the probability of success and failure have been scaled by a factor ϵ. The composed system is a mean field interaction model. This is due to the definition of functions tr_1 and tr_2, which do not depend on states of specific nodes, but rather on the aggregate number of nodes in states 1 and 2.

Consider the normalized population model with occupancy measures $\Delta = \{m \in \mathbb{R}^2 : \sum_i m_i = 1 \wedge \forall i. m_i \geq 0\}$ and the corresponding subset $\Delta^{(N)}$. Let $m \in \Delta^{(N)}$; then using (1) when the system is in state m the total number of communicating agents is $X_1 + X_2$ where $X_1 \sim B(Nm_1, p_1)$ and $X_2 \sim B(Nm_2, p_2)$. For an agent in state s the rate of moving to an state $s' \neq s$ is given by:

$$\begin{cases} Q_{1,2}^{(N)}(m) = \mathbb{E}\left[p1(1 - p_s(X_1 + X_2))\right], \\ Q_{2,1}^{(N)}(m) = \mathbb{E}\left[p_2 p_s(X_1 + X_2)\right]. \end{cases} \tag{3}$$

In the sections that follow, we use the map $Q^{(N)}$ to derive the mean field approximations of population processes.

3   Drift and the Time Evolution of Population Processes

In this section we define the drift as a way to characterize the behaviour of the population process $\bar{M}^{(N)}(t)$ in its first moment. This provides the basis for defining the mean drift, which is given in Sect. 4.

Define $W^{(N)}_{s,s'}(t)$ as the random number of objects which do a transition from state s to state s' in the system at time $t \in T_G$, i.e.,

$$W^{(N)}_{s,s'}(t + \epsilon) = \sum_{k=1}^{N} \mathbb{1}\left\{\hat{X}^{(N)}_k(t) = s, \hat{X}^{(N)}_k(t + \epsilon) = s'\right\}.$$

The instantaneous changes of the system $M^{(N)}(t)$ can be tracked by the following random process:

$$M^{(N)}(t + \epsilon) - M^{(N)}(t) = \sum_{s,s' \in S, s \neq s'} \frac{W^{(N)}_{s,s'}(t + \epsilon)}{N}(e_{s'} - e_s), \qquad (4)$$

where e_s is a unit vector of dimension I with value 1 in position s. Then the expected value of the instantaneous change is the function $\hat{F}^{(N)} : \Delta^{(N)} \to \mathbb{R}^I$ where:

$$\hat{F}^{(N)}(m) = \mathbb{E}\left[M^{(N)}(t + \epsilon) - M^{(N)}(t) \mid M^{(N)}(t) = m\right] = \sum_{s,s' \in S} m_s P^{(N)}_{s,s'}(m)(e_{s'} - e_s). \quad (5)$$

The *drift* is the function $F^{(N)} : \Delta^{(N)} \to \mathbb{R}^I$ defined as:

$$F^{(N)}(m) = \lim_{D \to \infty} D\hat{F}^{(N)}(m) = \sum_{s,s' \in S, s \neq s'} m_s Q^{(N)}_{s,s'}(m)(e_{s'} - e_s).$$

In the above formula, we may use $F^{(N)}_{s,s'}(m)$ to represent the summands $m_s Q^{(N)}_{s,s'}(m)$, which we call the *intensity* of transitions from s to s'. Essentially, the drift extends the vector representing the expected instantaneous changes into the unit time interval.

In the following sections we assume that the drifts of the systems we consider in our approximations satisfy the following assumptions.

Smoothness: For all $N \geq 1$, there exist Lipschitz continuous functions $\bar{F}^{(N)} : \Delta \to \mathbb{R}^I$ which for all $m \in \Delta^{(N)}$ satisfy: $\bar{F}^{(N)}(m) = F^{(N)}(m)$.

Boundedness: For all $N \geq 1$, $F^{(N)}$ are bounded on $\Delta^{(N)}$.

Limit existence: Assuming **smoothness**, the sequence of drifts $\{F^{(N)}\}$ converges uniformly to a bounded function $F^* : \Delta \to \mathbb{R}^I$.

In the literature, the single term *density dependence* is often used to refer to **boundedness** and **limit existence** [15].

Remark 1. In the contexts where it is clear that **smoothness** holds, we overload the name $F^{(N)}$ to refer to the function $\bar{F}^{(N)}$ instead.

Using drift, one can express how the expected value $\mathbb{E}[\bar{M}^{(N)}(t)]$ will evolve over time, a fact which we formally express through the following proposition.

Proposition 1. *For the process $\bar{M}^{(N)}(t)$, and its drift $F^{(N)}$ the following equation holds:*

$$\mathbb{E}\left[\bar{M}^{(N)}(t) \mid \bar{M}^{(N)}(0)\right] - \bar{M}^{(N)}(0) = \int_0^t \mathbb{E}\left[F^{(N)}(\bar{M}^{(N)}(s)) \mid \bar{M}^N(0)\right] ds.$$

Proof. A sketch of the proof follows [16]. For the identity function $f : \Delta \to \Delta$ the pair $(f, F^{(N)})$ belongs to the set A of infinitesimal generators of $\bar{M}^{(N)}(t)$. This then implies that the pair satisfies Dynkin's formula, as given in the equation above.   □

In its differential form, the equation above suggests that the expected trajectory of the process $\bar{M}^{(N)}(t)$ is a solution of the following system of ordinary differential equations:

$$\frac{d}{dt}\mathbb{E}\left[\bar{M}^{(N)}(t)\right] = \mathbb{E}\left[F^{(N)}(\bar{M}^{(N)}(t))\right], \tag{6}$$

with the initial value $\bar{M}^{(N)}(0)$. In practice the term $\mathbb{E}\left[F^{(N)}(\bar{M}^{(N)}(t))\right]$ is difficult to describe. In Sect. 4 we propose a way to approximate the right hand side of Eq. (6) by expressing it in terms of $\mathbb{E}\left[\bar{M}^{(N)}(t)\right]$, without explicitly giving the error bounds.

In the following example, we continue towards a mean field approximation for the system defined in example 1.

Example 2. We use the maps $Q_{s,s'}^{(N)}$ given by (3) to derive the drift of the system described in Example 1. A simple substitution gives the following sequence of drifts:

$$F^{(N)}(m) = \begin{pmatrix} -m_1\mathbb{E}\left[p1(1 - p_s(X_1 + X_2))\right] + m_2\mathbb{E}\left[p_2p_s(X_1 + X_2)\right] \\ -m_2\mathbb{E}\left[p_2p_s(X_1 + X_2)\right] + m_1\mathbb{E}\left[p1(1 - p_s(X_1 + X_2))\right] \end{pmatrix} \tag{7}$$

in which $X_1 \sim B(Nm_1, p_1)$ and $X_2 \sim B(Nm_2, p_2)$. This simplifies to:

$$F^{(N)}(m) = \begin{pmatrix} -m_1p_1(1 - (1 - \frac{p_1}{2})^{Nm_1}(1 - \frac{p_2}{2})^{Nm_2}) + m_2p_2(1 - \frac{p_1}{2})^{Nm_1}(1 - \frac{p_2}{2})^{Nm_2} \\ m_1p_1(1 - (1 - \frac{p_1}{2})^{Nm_1}(1 - \frac{p_2}{2})^{Nm_2}) - m_2p_2(1 - \frac{p_1}{2})^{Nm_1}(1 - \frac{p_2}{2})^{Nm_2} \end{pmatrix}$$

It can be shown that the inequality below is always satisfied for $m, m' \in \Delta$:

$$\left|F^{(N)}(m') - F^{(N)}(m)\right| \leq \sqrt{2}\,|m' - m|,$$

which proves that $F^{(N)}$ are Lipschitz continuous on Δ. In the same manner, it can be shown that for all $m \in \Delta$, $|F^{(N)}(m)| \leq \sqrt{2}$.

Therefore it is safe to assume that $F^{(N)}$ satisfies both **smoothness** and **boundedness**. Moreover, for any $p_1, p_2 > 0$ and $m \in \Delta$ we have:

$$F^*(m) = \lim_{N \to \infty} F^{(N)}(m) = \begin{pmatrix} -p_1m_1 \\ p_1m_1 \end{pmatrix} \tag{8}$$

which shows that **limit existence** is also satisfied.

Approximations of Mean Field Interaction Models

In this part, we explain how the drift satisfying all the discussed conditions (**smoothness, boundedness** and **limit existence**) can be used to derive a deterministic approximation of the behaviour of the population process.

For $N \geq 1$, let $F^{(N)}$ be a drift satisfying **smoothness**, and consider the following system of ordinary differential equations (ODEs):

$$\frac{d}{dt}\phi^{(N)}(t) = F^{(N)}(\phi^{(N)}(t)).$$

Take $\phi(0) = \phi_0$ as the initial condition of the above system of ODEs. Then based on the Picard-Lindelöf theorem since $F^{(N)}$ is Lipschitz continuous, a unique solution to the above system of ODEs exists. The unique function $\phi^{(N)}(t)$ which satisfies $\phi^{(N)}(0) = \phi_0$ is henceforth called the *deterministic approximation*.

Assuming that **limit existence** holds, define the *limit system of ODEs* as: $\phi'(t) = F^*(\phi(t))$ with initial condition ϕ_0. The solution $\phi(t)$ satisfies the following theorem.

Theorem 1 (*Mean Field Approximation, cf.* [2], *Theorem 1*). *For $N \geq 1$, let $\{\bar{M}^{(N)}(t)\}$ be a sequence of normalised population processes. Let $\{F^{(N)}\}$ be the corresponding drifts which satisfy* **smoothness**, **boundedness** *and* **limit existence**. *Let $\phi(t)$ be the solution to the corresponding limit system of ODE and assume $\lim_{N \to \infty} |\bar{M}^{(N)}(0) - \phi(0)|^2 = 0$. Then for any finite time horizon $T < \infty$:*

$$\lim_{N \to \infty} \mathbb{E}\left[\sup_{0 \leq t \leq T} \left|\bar{M}^{(N)}(t) - \phi(t)\right|^2\right] = 0.$$

Proof. This is the sketch of the proof in [16], which itself is an excerpt from a proof in [1]. Using a number of earlier results, it can be shown that the time evolution of the population process can be captured by a summation of unit Poisson processes, dependent on the drift. This is called the Poisson representation of the population process.

Next, these unit Poisson processes are decomposed into a summation of their expectation and compensated unit Poisson processes. Compensated Poisson processes are martingales with expectation 0. In this case, using Doob's inequality and **boundedness**, the maximum of the noise generated by these processes is shown to be bounded by a term of $O\left(\frac{T}{N}\right)$.

Finally, by using **smoothness** and applying Gröwall's inequality, it can be shown that for $T < \infty$ and at time $t \leq T$, as $N \to \infty$ the error in the approximation of $\bar{M}^{(N)}(t)$ by $\phi(t)$ almost surely tends to 0.   □

In the following example, we derive the limit system of ODEs for the system discussed in Example 2 and describe its solutions.

Example 3. Consider the drift given in (8). The limit system of ODEs for the system described in Example 1 is:

$$\frac{d}{dt}\phi_1(t) = -p_1\phi_1(t), \quad \frac{d}{dt}\phi_2(t) = p_1\phi_1(t).$$

which together with the initial condition $\phi(0)$ has the following general solution:

$$\phi_1(t) = \phi_1(0)\, e^{-p_1 t}, \quad \phi_2(t) = -\phi_1(0)\, e^{-p_1 t} + 1.$$

Obviously the solution heavily depends on the initial values $\phi(0)$, but the system has a global attractor at $(0, 1)$.

4   Propagation of Chaos and the Mean Drift

Theorem 1 justifies the use of drift for finding the approximation in cases where the number of agents N is unboundedly large. However, the bounds of error for the approximation (see [16]) barely justifies its use in the analysis of middle-sized systems. In this section we explore the possibility of using the alternative ODEs in (6). We explain the notion of propagation of chaos, and show how it relates to what we call the *mean drift*.

For a set E let $M(E)$ denote the set of probability measures on E. Let the set $\mathcal{S}$ be defined as in Sect. 2.1, and for $s, s' \in \mathcal{S}$ define the distance between s and s' as the function $d(s, s')$ which has value 2 if $s \neq s'$ and 0 otherwise. This makes the pair $(\mathcal{S}, d)$ a metric space, with the implication that $\mathcal{S}^N$ is also metrizable, an important condition which allows the definition that follows.

Definition 2 (*ρ-chaotic Sequence*). *Let $\rho \in M(\mathcal{S})$ be a probability measure. For $N \geq 1$, the sequence $\{\rho_N\}$ of measures, each in $M(\mathcal{S}^N)$, is ρ-chaotic iff for any fixed natural number k and bounded functions $f_1, \ldots, f_k : \mathcal{S} \to \mathbb{R}$,*

$$\lim_{N \to \infty} \int_{\mathcal{S}^N} f_1(v_1) f_2(v_2) \ldots f_k(v_k) \rho_N(dv) = \prod_{i=1}^{k} \int_{\mathcal{S}} f_i(s)\, \rho(ds).$$

In short, a chaotic sequence maintains a form of independence in the observations of separate agents in the limit. This independence is often called the *propagation of chaos* in literature, and *decoupling assumption* in the context of Bianchi's analysis.

In the following discussion, consider the instant in time $t \in \mathbb{R}_{\geq 0}$ and its close rational counterpart $\tau \in T_G$ with $\tau = \epsilon \lfloor Dt \rfloor$. Recall the non-normalized mean field interaction model at time τ, $Y^{(N)}(\tau) = \left(\hat{X}_1(\tau), \ldots, \hat{X}_N(\tau) \right)$, which is a random element in $\mathcal{S}^N$. For $N \geq 1$, let $\rho_N \in M(\mathcal{S}^N)$ be laws (probability distributions) of $Y^{(N)}(\tau)$. The following result shows that propagation of chaos occurs in the sequence of distributions of mean field interaction models $Y^{(N)}(\tau)$.

Corollary 1. *Let $\phi(t)$ satisfy Theorem 1. Let $\mu \in M(\mathcal{S})$ be a measure which for all points $i \in \mathcal{S}$ satisfies $\mu(i) = \phi_i(t)$, then the sequence $\{\rho_N\}$ of distributions of $Y^{(N)}(\tau)$ is μ-chaotic.*

Proof. Based on Theorem 1, it can be shown that the distribution of occupancy measures $\bar{M}_i^{(N)}(t)$ converge to a Dirac measure centred at $\phi(t)$ [16]. Following a result by Sznitman (see [17], Proposition 2.2) this then shows that the sequence $\{\rho_N\}$ is μ-chaotic. $\square$

Let the measure μ be defined as in Corollary 1. We show how the above result can be used by considering an example. The probability of agent 1 being in state i and agent 2 not being in state i in $Y^{(N)}(\tau)$, is as follows:

$$\mathbb{P}\left\{X_1^{(N)}(t) = i \wedge X_2^{(N)}(t) \neq i\right\} = \int_{S^N} \mathbb{1}(v_1 = i)\mathbb{1}(v_2 \neq i)\rho_N(dv).$$

Since $\{\rho_N\}$ is a μ-chaotic sequence, based on Definition 2:

$$\lim_{N \to \infty} \mathbb{P}\left\{X_1^{(N)}(t) = i \wedge X_2^{(N)}(t) \neq i\right\} = \phi_i(t)(1 - \phi_i(t)),$$

which means that the probability of finding the agents in the above configuration is asymptotically independent.

Next, for $k \in \{0, \ldots, N\}$, we are interested in finding the probability:

$$\mathbb{P}\left\{N\bar{M}_i^{(N)}(t) = k\right\},$$

given that the distributions of $\{Y^{(N)}(\tau)\}$ form a μ-chaotic sequence. Using the fact that $Y^N(\tau)$ is a mean field interaction model (is symmetric [17]), we have:

$$\mathbb{P}\left\{N\bar{M}_i^{(N)}(t) = k\right\} = \binom{N}{k} \int_{S^N} \mathbb{1}(v_1 = i) \ldots \mathbb{1}(v_k = i) \mathbb{1}(v_{k+1} \neq i) \ldots \mathbb{1}(v_N \neq i)\rho_N(dv).$$

It is worth noting that as $N \to \infty$, the number of functions considered on the right hand side tends to become infinite, and thus Definition 2 cannot be directly applied. However, this inspires us to propose the following approximation by assuming that all the agents in such a system become asymptotically independent:

$$\mathbb{P}\left\{N\bar{M}_i^{(N)}(t) = k\right\} \approx \binom{N}{k} \left(\phi_i^{(N)}(t)\right)^k \left(1 - \phi_i^{(N)}(t)\right)^{N-k},$$

or the following, which is often more easy to use:

$$\mathbb{P}\left\{N\bar{M}_i^{(N)}(t) = k\right\} \approx e^{-N\phi_i^{(N)}(t)} \frac{\left(N\phi_i^{(N)}(t)\right)^k}{k!}. \tag{9}$$

In the context of Bianchi's analysis, see [18] for an implicit application of a similar approximation.

Let $f_{Poisson}(k; \lambda)$ denote the probability density function of a Poisson random variable with rate λ. Let $f : \Delta \to \mathbb{R}$ be a function which acts on the random variable $\bar{M}^{(N)}(t)$. A major convenience in using the above terms is their asymptotic mutual independence, which at time $t \in \mathbb{R}_{\geq 0}$ allows the approximation of the expected value of $f(\bar{M}^{(N)}(t))$ as:

$$\mathbb{E}\left[f(\bar{M}^{(N)}(t))\right] \approx \sum_{k_1=0}^{\infty} \cdots \sum_{k_I=0}^{\infty} f\left(\frac{k_1}{N}, \ldots, \frac{k_I}{N}\right) f_{Poisson}\left(k_1; N\phi_1^{(N)}(t)\right) \ldots f_{Poisson}\left(k_I; N\phi_I^{(N)}(t)\right).$$

$$\tag{10}$$

The Mean Drift

In this part, we explain the approximation of the ODEs in formula (6), using result (10). For the occupancy measure $m \in \Delta$, the *Poisson mean* of the intensity $F_{s,s'}^{(N)}$ is the function $\tilde{F}_{s,s'}^{(N)} : \Delta \to \mathbb{R}$, where:

$$\tilde{F}_{s,s'}^{(N)}(m) = \sum_{k_1=0}^{\infty} \cdots \sum_{k_I=0}^{\infty} F_{s,s'}^{(N)}\left(\frac{k_1}{N}, \ldots, \frac{k_I}{N}\right) f_{Poisson}(k_1; Nm_1) \ldots f_{Poisson}(k_I; Nm_I). \quad (11)$$

Subsequently, the *mean drift* $\tilde{F}^{(N)} : \Delta \to \mathbb{R}^I$ is defined as:

$$\tilde{F}^{(N)}(m) = \sum_{s,s' \in \mathcal{S}} \tilde{F}_{s,s'}^{(N)}(m)(e_{s'} - e_s).$$

The Poisson mean of intensities and the mean drift have the following properties:

- If $F_{s,s'}^{(N)}(m) = m_s \alpha$, for α a constant or a term m_j for some $j \in \mathcal{S}$, then $\tilde{F}_{s,s'}^{(N)}(m) = F_{s,s'}^{(N)}(m)$.
- $\tilde{F}^{(N)}(m)$ is defined for all $m \in \Delta$.
- Given **smoothness** and **boundedness** of the drift, $\tilde{F}^{(N)}$ is both Lipschitz continuous and bounded on Δ.

Based on the derivation of probabilities (9), it is easy to see that at time $t < \infty$, we have $\mathbb{E}\left[F^{(N)}(\bar{M}^N(t))\right] \approx \tilde{F}^{(N)}\left(\phi^{(N)}(t)\right)$ and $\mathbb{E}\left[\bar{M}^N(t)\right] \approx \phi^{(N)}(t)$, according to which the following system of differential equations can be derived from (6):

$$\frac{d}{dt}\phi^{(N)}(t) = \tilde{F}^{(N)}\left(\phi^{(N)}(t)\right) \quad (12)$$

with the initial condition $\phi^{(N)}(0) = \bar{M}^{(N)}(0)$.

Our construction which is inspired by the notion of propagation of chaos, means that the differential equations (12) give better approximations as the system size N grows. This fact is demonstrated by the following theorem.

Theorem 2. *For $N \geq 1$, let $\left\{F^{(N)}\right\}$ be the sequence of drifts, and $\left\{\tilde{F}^{(N)}\right\}$ be the corresponding sequence of mean drifts. Assume that the drifts satisfy* **smoothness, boundedness**. *Then for all $m \in \Delta$,*

$$\lim_{N \to \infty} \tilde{F}^{(N)}(m) = F^*(m),$$

almost surely.

Proof. The following is a sketch of the proof given in [16]. Let $m \in \Delta$ be an occupancy measure. The application of the law of large numbers to the Poisson mean of intensities in $\tilde{F}^{(N)}(m)$ suggests that the probability mass of the product of the Poisson terms concentrates in an arbitrarily small neighbourhood of m in Δ, as $N \to \infty$. Since $F^{(N)}$ are Lipschitz continuous in Δ, this then implies that the sequence $\left\{\tilde{F}^{(N)}(m)\right\}$ almost surely converges to $F^*(m)$. □

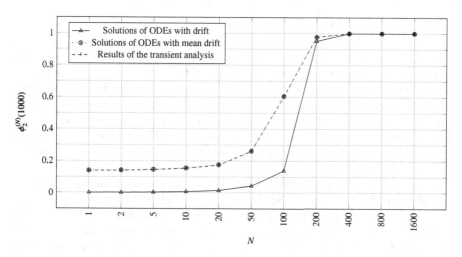

Fig. 2. Comparison between the proportion of nodes in the back-off state at time $t = 1000$ ($\phi_2^{(N)}(1000)$) for different network sizes N, based on a transient analysis of the Markov models (the solutions of the Chapman-Kolmogorov equations) and a mean field analysis by the ODEs incorporating the mean-drift.

The following is the final step in building an approximation using the mean drift for the system in Example 1, with the aim to partially demonstrate the accuracy of the proposed method of approximation.

Example 4. For the system described in Example 1 we find the mean drift, using the drift $F^{(N)}$ described in (7). The mean drift takes the relatively simple shape which follows:

$$\tilde{F}^{(N)}(\boldsymbol{m}) = \begin{pmatrix} \frac{p_1 m_1 + p_2 m_2}{2} exp\left\{-\frac{p_1 N m_1 + p_2 N m_2}{2}\right\} - p_1 m_1 \\ -\frac{p_1 m_1 + p_2 m_2}{2} exp\left\{-\frac{p_1 N m_1 + p_2 N m_2}{2}\right\} + p_1 m_1 \end{pmatrix}$$

This can be used to construct the following system of ODEs:

$$\frac{d}{dt}\phi_1^{(N)} = \frac{p_1 \phi_1^{(N)}(t) + p_2 \phi_2^{(N)}(t)}{2} exp\left\{-\frac{p_1 N \phi_1^{(N)}(t) + p_2 N \phi_2^{(N)}(t)}{2}\right\} - p_1 \phi_1^{(N)}(t)$$

$$\frac{d}{dt}\phi_2^{(N)} = -\frac{p_1 \phi_1^{(N)}(t) + p_2 \phi_2^{(N)}(t)}{2} exp\left\{-\frac{p_1 N \phi_1^{(N)}(t) + p_2 N \phi_2^{(N)}(t)}{2}\right\} + p_1 \phi_1^{(N)}(t)$$

Let $p_1 = 0.008$ and $p_2 = 0.05$ and let the initial condition be $\phi_0 = (1,0)$, i.e., all the nodes are initially in state 1. In Fig. 2 the results of solving the ODEs for the drift and the mean drift for different values of N are given, and are compared with results from the explicit transient analysis of Markov models with simultaneous updates. Observe that regardless of the size of the system, the approximations derived by the mean drift closely match the results of the transient analysis.

5   Conclusion and Future Work

In [19] the authors apply a version of the superposition principle called the *Poisson averaging* of the drift while deriving ODEs for wireless sensor networks. The method is used to cope with the ambiguous meaning of fractions which appear in arguments given to functions originally defined on discrete domains, and essentially interpolates the value of the function by interpreting occupancy measures as Poisson arrivals. In this paper we explain this practice more in detail by showing how they relate to other concepts in the mean field theory of Markov processes.

The result is the introduction of the mean drift, a concept that supports the analysis of bounded systems given that the mean field approximation applies. We maintain that within our formal framework the approximation theorems hold for systems with an infinite number of agents. We bridge the gap between the ODEs derived by using the mean drift and the ODEs describing the behaviour of the system in the limit by proving Theorem 2, which states that under a familiar set of conditions (smoothness and boundedness of the drift), the sequence of mean drifts converges to the limit of the sequence of drifts due to the law of large numbers.

We expect that for middle-sized systems, deriving ODEs using the mean drift gives far better approximations for their behaviour. As such, the current work provides a stepping stone for future efforts to apply the mean field analysis to the design and performance evaluation of distributed systems.

A natural next step to this work would be to provide bounds on the errors in the approximations found using the mean drift. That would then allow a more clear definition of middle-sized systems, and the type of behaviours which are better modelled by using this approach.

Acknowledgments. The research from DEWI project (www.dewi-project.eu) leading to these results has received funding from the ARTEMIS Joint Undertaking under grant agreement No. 621353.

References

1. Kurtz, T.G.: Approximation of Population Processes. CBMS-NSF Regional Conference Series in Applied Mathematics, vol. 36. SIAM, Philadelphia (1981)
2. Benaim, M., Le Boudec, J.-Y.: A class of mean field interaction models for computer and communication systems. Perform. Eval. **65**(11), 823–838 (2008)
3. Dietz, K., Heesterbeek, J.A.P.: Daniel Bernoulli's epidemiological model revisited. Math. Biosci. **180**(1), 1–21 (2002)
4. Darling, R.W.R., Norris, J.R.: Differential equation approximations for Markov chains. Probab. Surv. **5**, 37–79 (2008)
5. Kurtz, T.G.: Solutions of ordinary differential equations as limits of pure jump Markov processes. J. Appl. Probab. **7**(1), 49–58 (1970)
6. Ethier, S.N., Kurtz, T.G.: Markov Processes: Characterization and Convergence. Wiley, Hoboken (2009)

7. Bianchi, G.: Performance analysis of the IEEE 802.11 distributed coordination function. IEEE J. Sel. Areas Commun. **18**(3), 535–547 (2000)
8. Le Boudec, J.-Y., McDonald, D., Mundinger, J.: A generic mean field convergence result for systems of interacting objects. In: Fourth International Conference on the Quantitative Evaluation of Systems, QEST 2007, pp. 3–18. IEEE (2007)
9. Hillston, J.: Fluid flow approximation of PEPA models. In: QEST 2005, pp. 33–42. IEEE (2005)
10. Hayden, R.A., Bradley, J.T.: A fluid analysis framework for a Markovian process algebra. Theoret. Comput. Sci. **411**(22), 2260–2297 (2010)
11. Bortolussi, L., Hillston, J., Latella, D., Massink, M.: Continuous approximation of collective system behaviour: a tutorial. Perform. Eval. **70**(5), 317–349 (2013)
12. Pourranjbar, A., Hillston, J., Bortolussi, L.: Don't just go with the flow: cautionary tales of fluid flow approximation. In: Tribastone, M., Gilmore, S. (eds.) EPEW 2012. LNCS, vol. 7587, pp. 156–171. Springer, Heidelberg (2013). doi:10.1007/978-3-642-36781-6_11
13. Beccuti, M., Bibbona, E., Horvath, A., Sirovich, R., Angius, A., Balbo, G.: Analysis of Petri net models through Stochastic Differential Equations. In: Ciardo, G., Kindler, E. (eds.) PETRI NETS 2014. LNCS, vol. 8489, pp. 273–293. Springer, Cham (2014). doi:10.1007/978-3-319-07734-5_15
14. Bobbio, A., Gribaudo, M., Telek, M.: Analysis of large scale interacting systems by mean field method. In: Fifth International Conference on Quantitative Evaluation of Systems. QEST 2008, pp. 215–224. IEEE (2008)
15. Kurtz, T.G.: Strong approximation theorems for density dependent Markov chains. Stoch. Process. Their Appl. **6**(3), 223–240 (1978)
16. Talebi, M., Groote, J.F., Linnartz, J.-P.M.G.: The mean drift: tailoring the mean-field theory of Markov processes for real-world applications. arXiv preprint math/1703.04327 (2017)
17. Sznitman, A.-S.: Topics in propagation of chaos. In: Hennequin, P.-L. (ed.) Ecole d'Eté de Probabilités de Saint-Flour XIX — 1989. LNM, vol. 1464, pp. 165–251. Springer, Heidelberg (1991). doi:10.1007/BFb0085169
18. Vvedenskaya, N.D., Sukhov, Y.M.: Multiuser multiple-access system: stability and metastability. Problemy Peredachi Informatsii **43**(3), 105–111 (2007)
19. Talebi, M., Groote, J.F., Linnartz, J.-P.M.G.: Continuous approximation of stochastic models for wireless sensor networks. In: 2015 IEEE Symposium on Communications and Vehicular Technology in the Benelux (SCVT), pp. 1–6. IEEE (2015)

Author Index

Printed in the United States
By Bookmasters